Parametric Modeling

with Autodesk® Fusion 360®

Randy H. Shih

Oregon Institute of Technology

SDC
PUBLICATIONS

SDC Publications
P.O. Box 1334
Mission, KS 66222
913-262-2664
www.SDCpublications.com
Publisher: Stephen Schroff

ISBN-13: 978-1-63057-372-0
ISBN-10: 1-63057-372-8

Printed and bound in the United States of America.

Third Edition: March, 2020

Preface

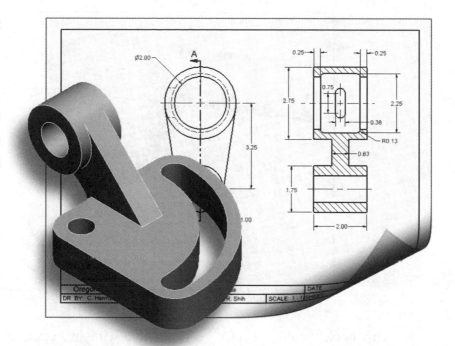

The primary goal of *Parametric Modeling with Autodesk Fusion 360* is to introduce the aspects of designing with **Solid Modeling** and **Parametric Modeling**. This text is intended to be used as a practical training guide for students and professionals. This text uses Autodesk Fusion 360 as the modeling tool and the chapters proceed in a pedagogical fashion to guide you from constructing basic solid models to building intelligent mechanical designs, creating multi-view drawings and assembly models. This text takes a hands-on, exercise-intensive approach to all the important *Parametric Modeling* techniques and concepts. This textbook contains a series of sixteen tutorial style lessons designed to introduce beginning CAD users to Autodesk Fusion 360. The solid modeling techniques and concepts discussed in this text are also applicable to other parametric feature-based CAD packages. The basic premise of this book is that the more designs you create using Autodesk Fusion 360, the better you learn the software. With this in mind, each lesson introduces a new set of commands and concepts, building on previous lessons. This book does not attempt to cover all of Autodesk Fusion 360's features, only to provide an introduction to the software. It is intended to help you establish a good basis for exploring and growing in the exciting field of **Computer Aided Engineering**.

Acknowledgments

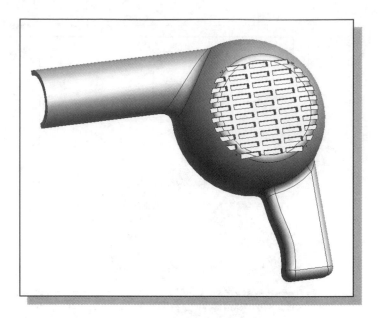

This book would not have been possible without a great deal of support. First, special thanks to two great teachers, Prof. George R. Schade of University of Nebraska-Lincoln and Mr. Denwu Lee, who showed me the fundamentals, the intrigue, and the sheer fun of Computer Aided Engineering.

The effort and support of the editorial and production staff of SDC Publications is gratefully acknowledged. I would especially like to thank Stephen Schroff for his support and helpful suggestions during this project.

I am grateful that the Department of Mechanical and Manufacturing Engineering at Oregon Institute of Technology has provided me with an excellent environment in which to pursue my interests in teaching and research.

Finally, truly unbounded thanks are due to my wife Hsiu-Ling and our daughter Casandra for their understanding and encouragement throughout this project.

Randy H. Shih
Klamath Falls, Oregon
Spring, 2020

Table of Contents

Chapter 3
Constructive Solid Geometry Concepts

Chapter 4
Model History Tree

Chapter 5
Parametric Constraints Fundamentals

Chapter 6
Geometric Construction Tools

Chapter 7
Parent/Child Relationships and the BORN Technique

Chapter 8
Part Drawings and Associative Functionality

Chapter 11
Symmetrical Features in Designs

Chapter 12
Advanced 3D Construction Tools

Chapter 13
Assembly Modeling - Joint & Animation

Index

Notes:

Chapter 1
Introduction – Getting Started

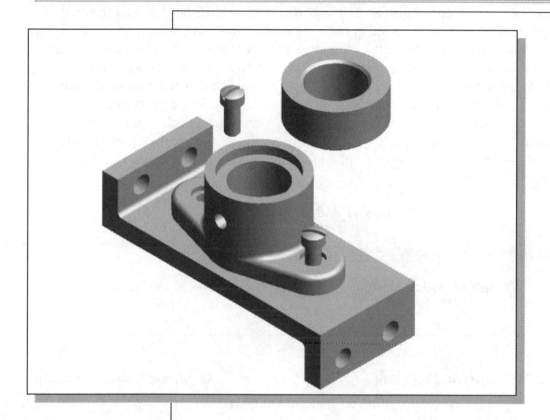

Learning Objectives

- ◆ **Development of Computer Geometric Modeling**
- ◆ **Feature-Based Parametric Modeling**
- ◆ **Startup Options and Units Setup**
- ◆ **Autodesk Fusion 360 Screen Layout**
- ◆ **User Interface & Mouse Buttons**
- ◆ **Autodesk Fusion 360 Online Help**

Introduction

The rapid changes in the field of **Computer Aided Engineering** (CAE) have brought exciting advances in the engineering community. Recent advances have made the long-sought goal of **concurrent engineering** closer to a reality. CAE has become the core of concurrent engineering and is aimed at reducing design time, producing prototypes faster, and achieving higher product quality. Autodesk Fusion 360 is a cloud based integrated package of mechanical computer aided engineering software tools developed by *Autodesk, Inc.* Autodesk Fusion 360 is a tool that facilitates a concurrent engineering approach to the design and stress-analysis of mechanical engineering products. The computer models can also be used by manufacturing equipment such as machining centers, lathes, mills, or rapid prototyping machines to manufacture the product. In this text, we will be dealing only with the solid modeling modules used for part design and part drawings.

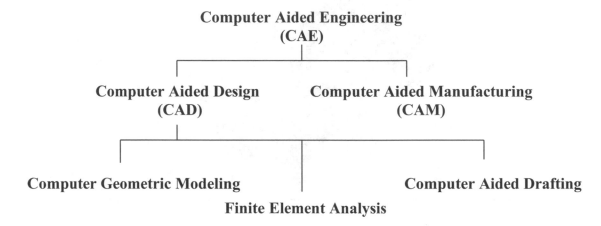

Development of Computer Geometric Modeling

Computer geometric modeling is a relatively new technology, and its rapid expansion in the last fifty years is truly amazing. Computer-modeling technology has advanced along with the development of computer hardware. The first-generation CAD programs, developed in the 1950s, were mostly non-interactive; CAD users were required to create program-codes to generate the desired two-dimensional (2D) geometric shapes. Initially, the development of CAD technology occurred mostly in academic research facilities. The Massachusetts Institute of Technology, Carnegie-Mellon University, and Cambridge University were the leading pioneers at that time. The interest in CAD technology spread quickly and several major industry companies, such as General Motors, Lockheed, McDonnell, IBM, and Ford Motor Co., participated in the development of interactive CAD programs in the 1960s. Usage of CAD systems was primarily in the automotive industry, aerospace industry, and government agencies that developed their own programs for their specific needs. The 1960s also marked the beginning of the development of finite element analysis methods for computer stress analysis and computer aided manufacturing for generating machine tool paths.

The 1970s are generally viewed as the years of the most significant progress in the development of computer hardware, namely the invention and development of **microprocessors**. With the improvement in computing power, new types of 3D CAD programs that were user-friendly and interactive became reality. CAD technology quickly expanded from very simple **computer aided drafting** to very complex **computer aided design**. The use of 2D and 3D wireframe modelers was accepted as the leading-edge technology that could increase productivity in industry. The developments of surface modeling and solid modeling technologies were taking shape by the late 1970s, but the high cost of computer hardware and programming slowed the development of such technology. During this period, the available CAD systems all required room-sized mainframe computers that were extremely expensive.

In the 1980s, improvements in computer hardware brought the power of mainframes to the desktop at less cost and with more accessibility to the general public. By the mid-1980s, CAD technology had become the main focus of a variety of manufacturing industries and was very competitive with traditional design/drafting methods. It was during this period of time that 3D solid modeling technology had major advancements, which boosted the usage of CAE technology in industry.

The introduction of the *feature-based parametric solid modeling* approach, at the end of the 1980s, elevated CAD/CAM/CAE technology to a new level. In the 1990s, CAD programs evolved into powerful design/manufacturing/management tools. CAD technology has come a long way, and during these years of development, modeling schemes progressed from two-dimensional (2D) wireframe to three-dimensional (3D) wireframe, to surface modeling, to solid modeling and, finally, to feature-based parametric solid modeling.

The first-generation CAD packages were simply 2D **computer aided drafting** programs, basically the electronic equivalents of the drafting board. For typical models, the use of this type of program would require that several views of the objects be created individually as they would be on the drafting board. The 3D designs remained in the designer's mind, not in the computer database. Mental translations of 3D objects to 2D views are required throughout the use of these packages. Although such systems have some advantages over traditional board drafting, they are still tedious and labor intensive. The need for the development of 3D modelers came quite naturally, given the limitations of the 2D drafting packages.

The development of three-dimensional modeling schemes started with three-dimensional (3D) wireframes. Wireframe models are models consisting of points and edges, which are straight lines connecting between appropriate points. The edges of wireframe models are used, similar to lines in 2D drawings, to represent transitions of surfaces and features. The use of lines and points is also a very economical way to represent 3D designs.

The development of the 3D wireframe modeler was a major leap in the area of computer geometric modeling. The computer database in the 3D wireframe modeler contains the locations of all the points in space coordinates, and it is typically sufficient to create just one model rather than multiple views of the same model. This single 3D model can then be viewed from any direction as needed. Most 3D wireframe modelers allow the user to create projected lines/edges of 3D wireframe models. In comparison to other types of 3D modelers, the 3D wireframe modelers require very little computing power and generally can be used to achieve reasonably good representations of 3D models. However, because surface definition is not part of a wireframe model, all wireframe images have the inherent problem of ambiguity. Two examples of such ambiguity are illustrated.

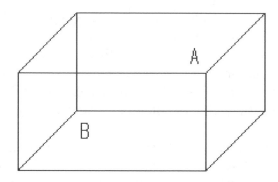

Wireframe Ambiguity: Which corner is in front, A or B?

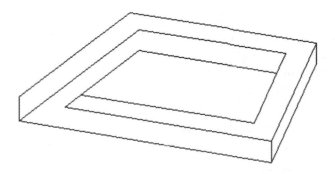

A non-realizable object: Wireframe models contain no surface definitions.

Surface modeling is the logical development in computer geometry modeling to follow the 3D wireframe modeling scheme by organizing and grouping edges that define polygonal surfaces. Surface modeling describes the part's surfaces but not its interiors. Designers are still required to interactively examine surface models to ensure that the various surfaces on a model are contiguous throughout. Many of the concepts used in 3D wireframe and surface modelers are incorporated in the solid modeling scheme, but it is solid modeling that offers the most advantages as a design tool.

In the solid modeling presentation scheme, the solid definitions include nodes, edges, and surfaces, and it is a complete and unambiguous mathematical representation of a precisely enclosed and filled volume. Unlike the surface modeling method, solid modelers start with a solid or use topology rules to guarantee that all of the surfaces are stitched together properly. Two predominant methods for representing solid models are **constructive solid geometry** (CSG) representation and **boundary representation** (B-rep).

The CSG representation method can be defined as the combination of 3D solid primitives. What constitutes a "primitive" varies somewhat with the software but typically includes a rectangular prism, a cylinder, a cone, a wedge, and a sphere. Most solid modelers also allow the user to define additional primitives, which are shapes typically formed by the basic shapes. The underlying concept of the CSG representation method is very straightforward; we simply **add** or **subtract** one primitive from another. The CSG approach is also known as the machinist's approach, as it can be used to simulate the manufacturing procedures for creating the 3D object.

In the B-rep representation method, objects are represented in terms of their spatial boundaries. This method defines the points, edges, and surfaces of a volume, and/or issues commands that sweep or rotate a defined face into a third dimension to form a solid. The object is then made up of the unions of these surfaces that completely and precisely enclose a volume.

By the 1980s, a new paradigm called *concurrent engineering* had emerged. With concurrent engineering, designers, design engineers, analysts, manufacturing engineers, and management engineers all work together closely right from the initial stages of the design. In this way, all aspects of the design can be evaluated, and any potential problems can be identified right from the start and throughout the design process. Using the principles of concurrent engineering, a new type of computer modeling technique appeared. The technique is known as the *feature-based parametric modeling technique*. The key advantage of the *feature-based parametric modeling technique* is its capability to produce very flexible designs. Changes can be made easily, and design alternatives can be evaluated with minimum effort. Various software packages offer different approaches to feature-based parametric modeling, yet the end result is a flexible design defined by its design variables and parametric features.

Feature-Based Parametric Modeling

One of the key elements in the Autodesk Fusion 360 solid modeling software is its use of the **feature-based parametric modeling technique**. The feature-based parametric modeling approach has elevated solid modeling technology to the level of a very powerful design tool. Parametric modeling automates the design and revision procedures by the use of parametric features. Parametric features control the model geometry by the use of design variables. The word *parametric* means that the geometric definitions of the design, such as dimensions, can be varied at any time during the design process. Features are predefined parts or construction tools for which users define the key parameters. A part is described as a sequence of engineering features, which can be modified and/or changed at any time. The concept of parametric features makes modeling more closely match the actual design-manufacturing process than the mathematics of a solid modeling program. In parametric modeling, models and drawings are updated automatically when the design is refined.

Parametric modeling offers many benefits:

* **We begin with simple, conceptual models with minimal detail; this approach conforms to the design philosophy of "shape before size."**

* **Geometric constraints, dimensional constraints, and relational parametric equations can be used to capture design intent.**

* **The ability to update an entire system, including parts, assemblies and drawings, after changing one parameter of complex designs.**

* **We can quickly explore and evaluate different design variations and alternatives to determine the best design.**

* **Existing design data can be reused to create new designs.**

* **Quick design turn-around.**

One of the key features of Autodesk Fusion 360 is the use of an assembly-centric paradigm, which enables users to concentrate on the design without depending on the associated parameters or constraints. Users can specify how parts fit together and the Autodesk Fusion 360 *assembly-based fit function* automatically determines the parts' sizes and positions. This unique approach is known as the **Direct Adaptive Assembly approach**, which defines part relationships directly with no order dependency.

The *Adaptive Assembly approach* is a unique design methodology that can only be found in Autodesk Fusion 360. The goal of this methodology is to improve the design process and allows you, the designer, to **Design the Way You Think**.

Getting Started with Autodesk Fusion 360

Autodesk Fusion 360 is composed of several application software modules (these modules are called *applications*), all sharing a common database. In this text, the main concentration is placed on the solid modeling modules used for part design. The general procedures required in creating solid models, engineering drawings, and assemblies are illustrated.

How to start Autodesk Fusion 360 depends on the type of workstation and the particular software configuration you are using. With most *Windows* systems, you may select **Autodesk Fusion 360** on the *Start* menu or select the **Autodesk Fusion 360** icon on the desktop. Consult your instructor or technical support personnel if you have difficulty starting the software.

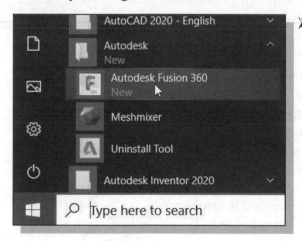

➢ If you already have an Autodesk account, log in with your email or username.

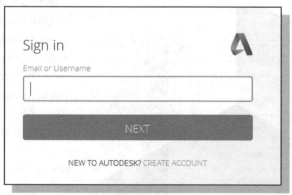

➢ To register a new account click **Create Account** and proceed to register a new account.

The Autodesk Fusion 360 Screen Layout

The default Autodesk Fusion 360 drawing screen contains the *Quick access* toolbar, the *Modeling* toolbar, the *display control* toolbar, the *Model Browser*, the *drawing* area, and the *View Cube*. You may resize the Autodesk Fusion 360 drawing window by clicking and dragging the edges of the window, or relocate the window by clicking and dragging the window title area.

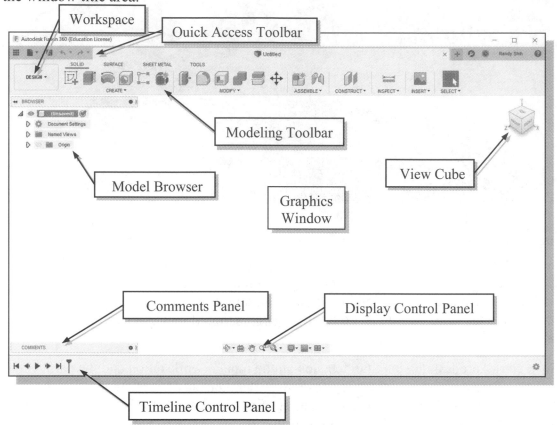

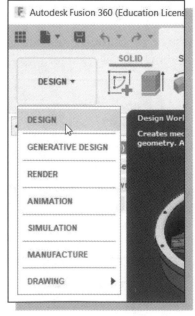

➢ Autodesk Fusion 360 is an integrated system that can be used to create parts, assembly, animation, simulation, CAM and 2D drawings. Select the appropriate **WorkSpace** for the desired task:

- **Design, Generative Design workspaces:** These are for creating solid, sheet metal and surface models.
- **Render workspace:** This workspace is for creating realistic images of designs.
- **Animation workspace:** This workspace is for assembling and performing motion analyses of designs.
- **Simulation workspace:** This workspace is for testing designs under different loadings.
- **Manufacture workspace:** This workspace is for generating tool-path to manufacture designs.
- **Drawing workspace:** This workspace is for creating drawings from parts and assemblies.

- **Quick Access Toolbar**

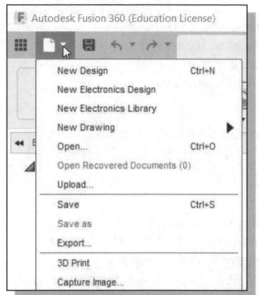

The *Quick Access* toolbar at the top of the *Fusion 360* window allows us quick access to file-related commands and to Undo/Redo the last operations.

- **Ribbon tool panels**

Depending on the selected workspace, the *Tool panels* display commands for the desired tasks.

- **Display Control Panel**

The *display control* panel provides us with multiple options to adjust the display of graphics area.

- **View Cube**

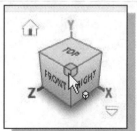

The *View Cube* provides an easy option to set the viewing direction of the design.

- **Comments Panel**

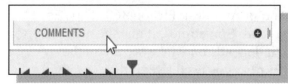

The *Comments panel* allows notes and comments to be attached to the designs, which can provide important information regarding the designs.

Mouse Buttons

Autodesk Fusion 360 utilizes the mouse buttons extensively. In learning Autodesk Fusion 360's interactive environment, it is important to understand the basic functions of the mouse buttons. It is highly recommended that you use a mouse or a tablet with Autodesk Fusion 360 since the package uses the buttons for various functions.

- **Left mouse button**
  The **left-mouse-button** is used for most operations, such as selecting menus and icons, or picking graphic entities. One click of the button is used to select icons, menus and form entries, and to pick graphic items.

- **Right mouse button**
  The **right-mouse-button** is used to bring up additional available options. The software also utilizes the **right-mouse-button** the same as the **ENTER** key and is often used to accept the default setting to a prompt or to end a process.

- **Middle mouse button/wheel**
  The middle mouse button/wheel can be used to **Rotate** (hold down the shift key and the wheel button while dragging the mouse), pan (hold down the wheel button and drag the the mouse) or **Zoom** (turn the wheel) realtime.

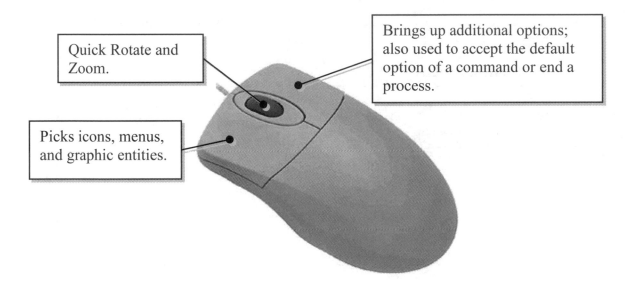

Quick Rotate and Zoom.

Brings up additional options; also used to accept the default option of a command or end a process.

Picks icons, menus, and graphic entities.

[Esc] – Canceling Commands

The [**Esc**] key is used to cancel or end a command in Autodesk Fusion 360. The [**Esc**] key is located near the top-left corner of the keyboard. Sometimes, it may be necessary to press the [**Esc**] key twice to cancel a command; it depends on where we are in the command sequence. For some commands, the [**Esc**] key is used to exit the command.

Autodesk Fusion 360 Help System

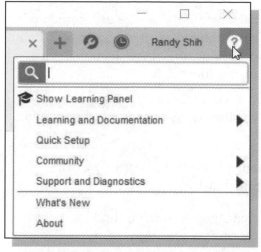

❖ Several types of help are available at any time during an Autodesk Fusion 360 session. Autodesk Fusion 360 provides many help functions.

- Click the **Help** button near the upper right corner of the *Fusion 360* window. Note the different available help options.

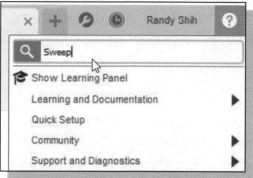

- Enter a keyword in the **Search option** to get information on a specific topic.

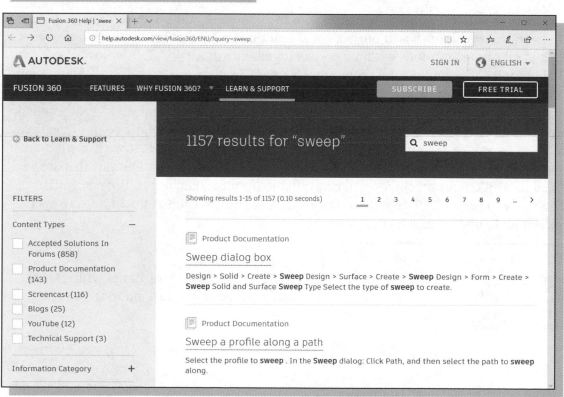

Autodesk Fusion 360 Data Management

Autodesk Fusion 360 is a cloud based integrated package; all of the files are stored at the central server and can be accessed anywhere. The advantages of using a cloud-based system are the ability of real time collaboration and working anywhere, on any device, and there is no need to worry about file storage or losing files.

Autodesk Fusion 360's integrated data management and collaboration platform enables the management of design data; we can also use it to share designs and collaborate with anyone in the world.

The Autodesk Fusion 360's integrated data management and collaboration platform allows us to perform the following project management tasks:

- Work with projects inside of Fusion 360 to manage file versions and invite team members to join the projects.

- Activate the Autodesk A360 project collaboration tool to allow collaborating with project teams and share design data inside and outside of the organization.

- Switch on the Autodesk A360 model viewer to comment on designs with the project team and stakeholders.

Autodesk Fusion 360 provides a fairly flexible data management system. The Autodesk Fusion 360 data management system organizes files based on **projects**. Each project is identified with a main folder that can contain files and folders associated with the design.

Autodesk Fusion 360 can also interface with the Autodesk A360 service. The Autodesk A360 service is a cloud-based project collaboration service that brings the people, the data and all of the activities within a project together. Autodesk A360 project collaboration software, which is run through a web browser or a mobile app, helps teams manage product data and work together on a centralized platform. We can view, share, and review both 2D and 3D designs in a central location. Using Autodesk A360 can help keep our projects, files, and teams up to date, whether we are at the office or in the field. Note that Autodesk Fusion 360 can also work in **offline mode**, so that we can still continue to work on our designs even without internet connections. Designs that are cached locally on our machines will be available from the data panel. We can work offline up to a period of **two weeks** if needed. After this, Fusion 360 will need to sync back online to update to the latest version.

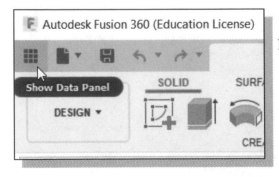

1. Select the **Show Data Panel** with a single click of the left-mouse button.

2. In the *Data Panel*, click on the **New Project** icon to create a new project.

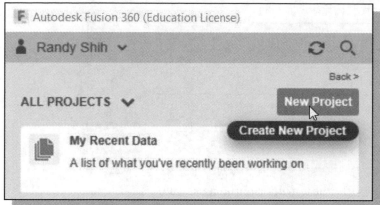

3. Enter **Parametric Modeling** as the new project name as shown.

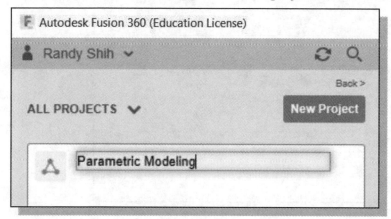

4. Click on the **Pin** icon to indicate this project is currently being worked on. Note that the project is moved up near the top of the data panel. We can pin any project as well as other projects we belong to. If we no longer want a project under the PINNED filter, we can unpin it.

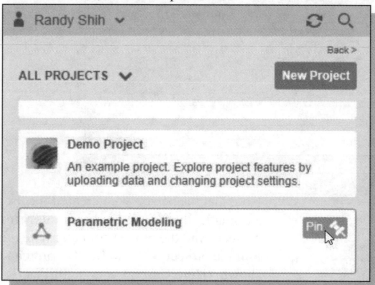

5. **Double click** on the project to view additional available project options.

6. In the *Parametric Modeling* project, click on the **New Folder** icon to create a new folder under the project.

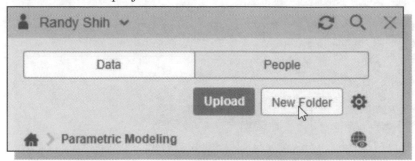

7. Enter **Chapter 1** as the new folder name as shown.

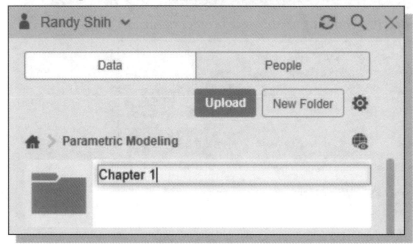

8. Click on the **People** icon to view the available team member options.

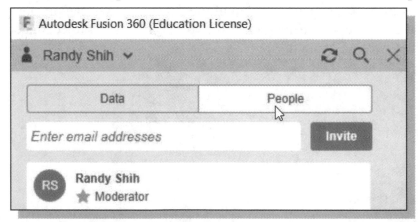

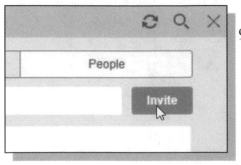

9. Note that the person who created the project is automatically assigned as the **moderator** of the project. And the moderator can invite others to join the project with the **Invite** command as shown. (Hint: Current version of Fusion 360 requires the email address contains only lower-case charters.)

10. In the Fusion 360 main window, click **Close Data Panel** to exit the *Data Panel* as shown.

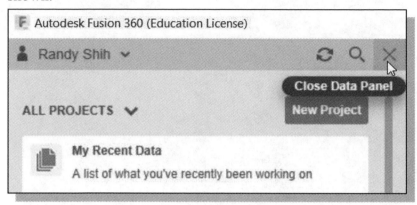

11. To exit *Autodesk Fusion 360*, click on **Close** in the upper right corner of the main window as shown.

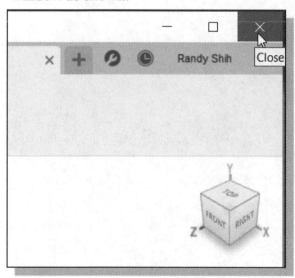

Notes:

Chapter 2
Parametric Modeling Fundamentals

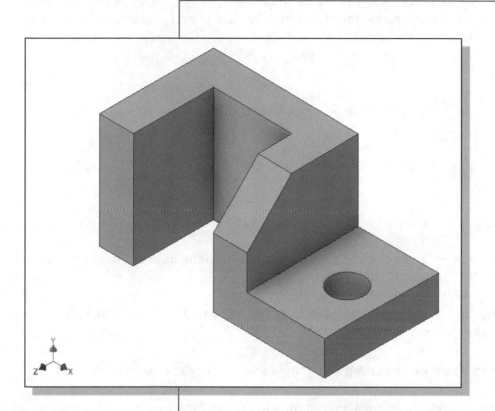

Learning Objectives

- ♦ **Create Simple Extruded Solid Models**
- ♦ **Understand the Basic Parametric Modeling Procedure**
- ♦ **Create 2D Sketches**
- ♦ **Understand the "Shape before Size" Design Approach**
- ♦ **Use the Dynamic Viewing Commands**
- ♦ **Create and Edit Parametric Dimensions**

Introduction

The **feature-based parametric modeling** technique enables the designer to incorporate the original **design intent** into the construction of the model. The word *parametric* means the geometric definitions of the design, such as dimensions, can be varied at any time in the design process. Parametric modeling is accomplished by identifying and creating the key features of the design with the aid of computer software. The design variables, described in the sketches as parametric relations, can then be used to quickly modify/update the design.

In Autodesk Fusion 360, the parametric part modeling process involves the following steps:

1. **Create a rough two-dimensional sketch of the basic shape of the base feature of the design.**

2. **Apply/modify constraints and dimensions to the two-dimensional sketch.**

3. **Extrude, revolve, or sweep the parametric two-dimensional sketch to create the base solid feature of the design.**

4. **Add additional parametric features by identifying feature relations and complete the design.**

5. **Perform analyses on the computer model and refine the design as needed.**

6. **Create the desired drawing views to document the design.**

The approach of creating two-dimensional sketches of the three-dimensional features is an effective way to construct solid models. Many designs are in fact the same shape in one direction. Computer input and output devices we use today are largely two-dimensional in nature, which makes this modeling technique quite practical. This method also conforms to the design process that helps the designer with conceptual design along with the capability to capture the ***design intent***. Most engineers and designers can relate to the experience of making rough sketches on restaurant napkins to convey conceptual design ideas. Autodesk Fusion 360 provides many powerful modeling and design-tools, and there are many different approaches to accomplishing modeling tasks. The basic principle of **feature-based modeling** is to build models by adding simple features one at a time. In this chapter, the general parametric part modeling procedure is illustrated; a very simple solid model with extruded features is used to introduce the Autodesk Fusion 360 user interface. The display viewing functions and the basic two-dimensional sketching tools are also demonstrated.

The Adjuster Design

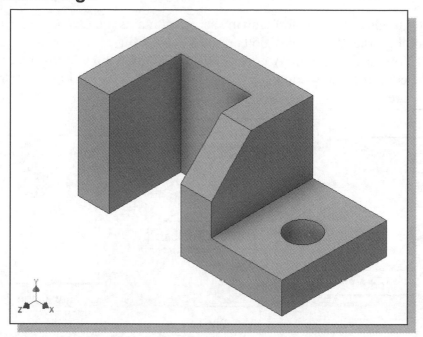

Starting Autodesk Fusion 360

1. Select the **Autodesk Fusion 360** option on the *Start* menu or select the **Autodesk Fusion 360** icon on the desktop to start Autodesk Fusion 360. The Autodesk Fusion 360 *log in* window will appear on the screen.

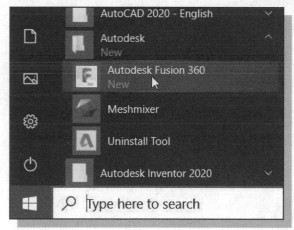

2. In the *Sign In* dialog box, log in with your email or username.

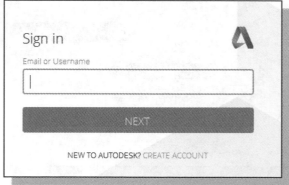

The Autodesk Fusion 360 Screen Layout

The default Autodesk Fusion 360 drawing screen contains the *Quick access* toolbar, the *Modeling* toolbar, the *Navigation* toolbar, the *Model Browser*, the *drawing* area, and the *View Cube*. You may resize the Autodesk Fusion 360 drawing window by clicking and dragging the edges of the window, or relocate the window by clicking and dragging the window title area.

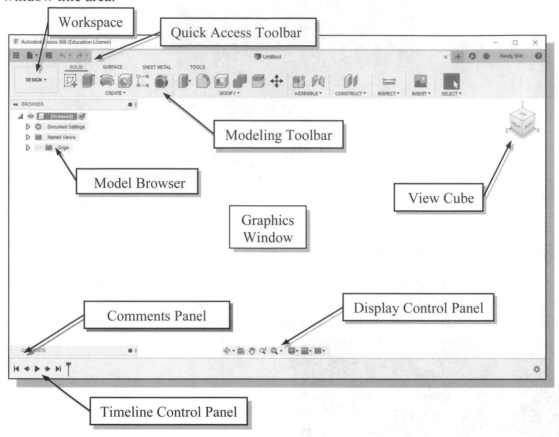

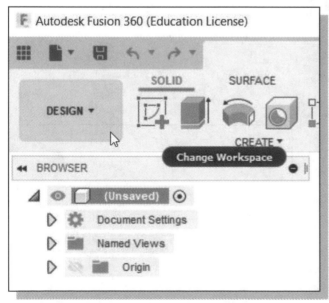

> Autodesk Fusion 360 is an integrated system that can be used to create parts, assembly, animation, simulation, CAM and 2D drawings. By default, the **Design WorkSpace** is activated as shown.

Units Setup

❖ Every object we construct in a CAD system is measured in units. We should determine the value of the units within the CAD system before creating the first geometric entities. For example, in one model, a unit might equal one millimeter of the real-world object; in another model, a unit might equal an inch. In Autodesk Fusion 360, the **_Document Settings_** option can be used to quickly display and also switch to using a different units set.

1. Click on the triangle icon next to the **_Document Settings_** option in the Browser area to show the default units as shown. Note the current units is set to the millimeter (mm) units as shown.

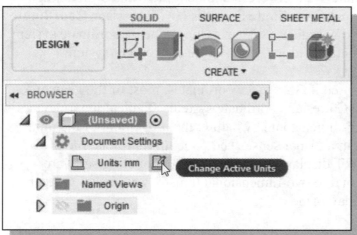

2. Click on the **_Change Active Units_** option in the Browser area as shown. We will use the inch (in) units for this example.

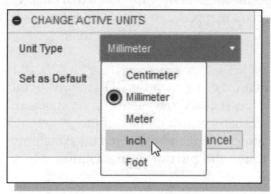

3. In the _Units_ list, select **Inch** as shown.

4. Pick **OK** to exit the _Change Active Units_ dialog box.

Sketch Plane – It is an XY monitor, but an XYZ World

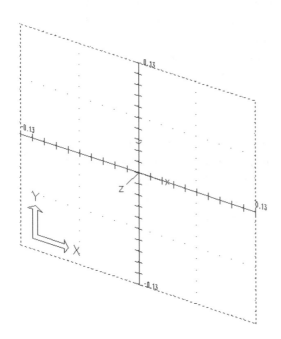

Design modeling software is becoming more powerful and user friendly, yet the system still does only what the user tells it to do. When using a geometric modeler, we therefore need to have a good understanding of what its inherent limitations are.

In most 3D geometric modelers, 3D objects are located and defined in what is usually called **world space** or **global space**. Although a number of different coordinate systems can be used to create and manipulate objects in a 3D modeling system, the objects are typically defined and stored using the world space. The world space is usually a **3D Cartesian coordinate system** that the user cannot change or manipulate.

In engineering designs, models can be very complex, and it would be tedious and confusing if only the world coordinate system were available. Practical 3D modeling systems allow the user to define **Local Coordinate Systems (LCS)** or **User Coordinate Systems (UCS)** relative to the world coordinate system. Once a local coordinate system is defined, we can then create geometry in terms of this more convenient system.

Although objects are created and stored in 3D space coordinates, most of the geometric entities can be referenced using 2D Cartesian coordinate systems. Typical input devices such as a mouse or digitizer are two-dimensional by nature; the movement of the input device is interpreted by the system in a planar sense. The same limitation is true of common output devices, such as CRT displays and plotters. The modeling software performs a series of three-dimensional to two-dimensional transformations to correctly project 3D objects onto the 2D display plane.

The Autodesk Fusion 360 *sketching plane* is a special construction approach that enables the planar nature of the 2D input devices to be directly mapped into the 3D coordinate system. The *sketching plane* is a local coordinate system that can be aligned to an existing face of a part, or a reference plane.

Think of the sketching plane as the surface on which we can sketch the 2D sections of the parts. It is similar to a piece of paper, a white board, or a chalkboard that can be attached to any planar surface. The first sketch we create is usually drawn on one of the established datum planes. Subsequent sketches/features can then be created on sketching planes that are aligned to existing **planar faces of the solid part** or **datum planes.**

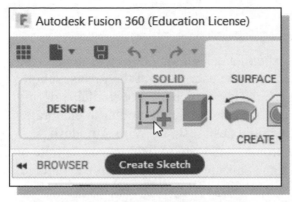

1. Activate the **Create Sketch** icon with a single click of the left-mouse-button.

2. Move the cursor over the right vertical plane, the *XY Plane,* in the graphics area. When the *XY Plane* is highlighted, click once with the **left-mouse-button** to select the *Plane* as the sketching plane for the new sketch.

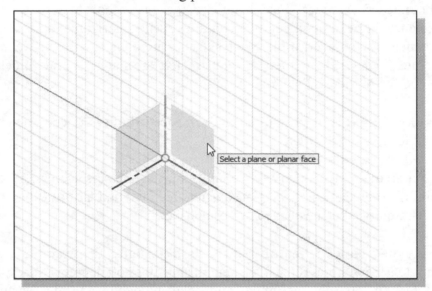

- The *sketching plane* is a reference location where two-dimensional sketches are created. Note that the *sketching plane* can be any planar part surface or datum plane.

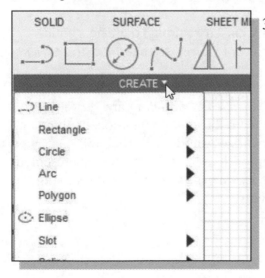

3. Click and press on the **Sketch** tab in the main *Ribbon* area to display the additional available sketching tools as shown. Also note the **Sketch Palette** panel is also activated with a varity of sketching options.

Creating Rough Sketches

Quite often during the early design stage, the shape of a design may not have any precise dimensions. Most conventional CAD systems require the user to input the precise lengths and locations of all geometric entities defining the design, which are not available during the early design stage. With *parametric modeling*, we can use the computer to elaborate and formulate the design idea further during the initial design stage. With Autodesk Fusion 360, we can use the computer as an electronic sketchpad to help us concentrate on the formulation of forms and shapes for the design. This approach is the main advantage of *parametric modeling* over conventional solid-modeling techniques.

As the name implies, a **rough sketch** is not precise at all. When sketching, we simply sketch the geometry so that it closely resembles the desired shape. Precise scale or lengths are not needed. Autodesk Fusion 360 provides many tools to assist us in finalizing sketches. For example, geometric entities such as horizontal and vertical lines are set automatically. However, if the rough sketches are poor, it will require much more work to generate the desired parametric sketches. Here are some general guidelines for creating sketches in Autodesk Fusion 360:

- **Create a sketch that is proportional to the desired shape.** Concentrate on the shapes and forms of the design.

- **Keep the sketches simple.** Leave out small geometry features such as fillets, rounds and chamfers. They can easily be placed using the Fillet and Chamfer commands after the parametric sketches have been established.

- **Exaggerate the geometric features of the desired shape.** For example, if the desired angle is 85 degrees, create an angle that is 50 or 60 degrees. Otherwise, Autodesk Fusion 360 might assume the intended angle to be a 90-degree angle.

- **Draw the geometry so that it does not overlap.** The geometry should eventually form a closed region. *Self-intersecting* geometry shapes are not allowed.

- **The sketched geometric entities should form a closed region.** To create a solid feature, such as an extruded solid, a closed region is required so that the extruded solid forms a 3D volume.

- ➢ **Note:** The concepts and principles involved in *parametric modeling* are very different, and sometimes they are totally opposite, to those of conventional computer aided drafting. In order to understand and fully utilize Autodesk Fusion 360's functionality, it will be helpful to take a *Zen* approach to learning the topics presented in this text: **Have an open mind and temporarily forget your experiences using conventional Computer Aided Drafting systems.**

Step 1: Creating a Rough Sketch

The *Sketch* toolbar provides tools for creating the basic geometry that can be used to create features and parts.

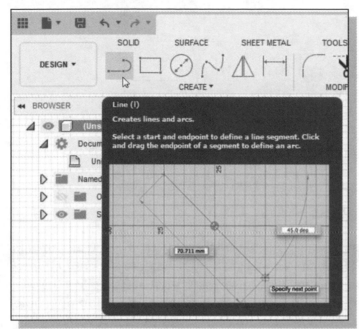

1. Move the graphics cursor to the **Line** icon in the *Sketch* toolbar. A *Help-tip box* appears next to the cursor and a brief description of the command is displayed at the bottom of the drawing screen: "*Creates lines and arcs.*"

2. Select the icon by clicking once with the **left-mouse-button**; this will activate the **Line** command. *Autodesk Fusion 360* expects us to identify the starting location of a straight line.

Graphics Cursors

Notice a crosshair cursor is shown when graphical input is expected.

1. Left-click a starting point for the shape, roughly just below the center of the graphics window.

2. As you move the graphics cursor, you will see a digital readout next to the cursor and also in the *Status Bar* area at the bottom of the window. The readout gives you the cursor location, the line length, and the angle of the line measured from horizontal. Move the cursor around and you will notice different symbols appear at different locations.

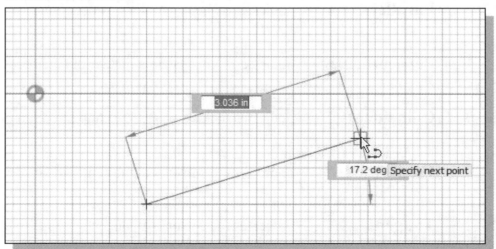

3. Move the graphics cursor toward the right side of the graphics window and create a horizontal line as shown below (**Point 2**). Notice the geometric constraint symbol, a short horizontal line indicating the geometric property, is displayed.

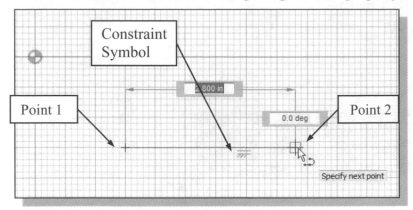

Geometric Constraint Symbols

Autodesk Fusion 360 displays different visual clues, or symbols, to show you alignments, perpendicularities, tangencies, etc. These constraints are used to capture the *design intent* by creating constraints where they are recognized. Autodesk Fusion 360 displays the governing geometric rules as models are built. To prevent constraints from forming, hold down the [**Ctrl**] key while creating an individual sketch curve. For example, while sketching line segments with the Line command, endpoints are joined with a Coincident *constraint*, but when the [**Ctrl**] key is pressed and held, the inferred constraint will not be created.

	Vertical	indicates a line is vertical
	Horizontal	indicates a line is horizontal
	Dashed line	indicates the alignment is to the center point or endpoint of an entity
	Parallel	indicates a line is parallel to other entities
	Perpendicular	indicates a line is perpendicular to other entities
	Coincident	indicates the cursor is at the endpoint of an entity
	Concentric	indicates the cursor is at the center of an entity
	Tangent	indicates the cursor is at tangency points to curves

1. Complete the sketch as shown below, creating a closed region ending at the starting point (**Point 1**). Do not be overly concerned with the actual size of the sketch. Note that all line segments are sketched horizontally or vertically.

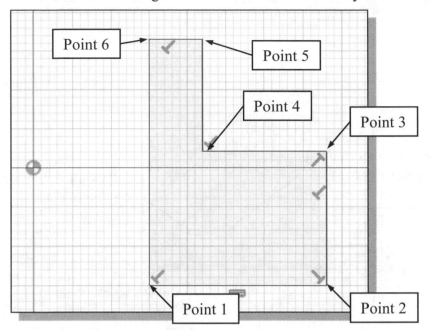

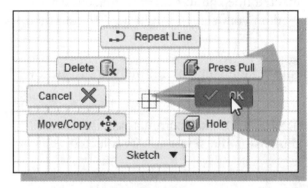

2. Inside the graphics window, click once with the **right-mouse-button** to display the option menu. Select [**OK**] in the pop-up menu, or hit the [**Esc**] key once to end the Sketch Line command.

Step 2: Apply/Modify Constraints and Dimensions

As the sketch is made, Autodesk Fusion 360 automatically applies some of the geometric constraints (such as horizontal, parallel, and perpendicular) to the sketched geometry. We can continue to modify the geometry, apply additional constraints, and/or define the size of the existing geometry. In this example, we will illustrate adding dimensions to describe the sketched entities.

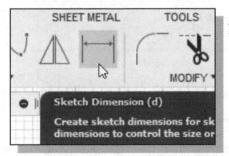

1. In the sketch toolbar, select the **Sketch Dimension** command as shown. The Sketch Dimension command is generally known as *Smart Dimensioning* in parametric modeling.

2. The message "*Select Geometry to Dimension*" is displayed in the *Status Bar* area at the bottom of the *Fusion 360* window. Select the bottom horizontal line by left-clicking once on the line.

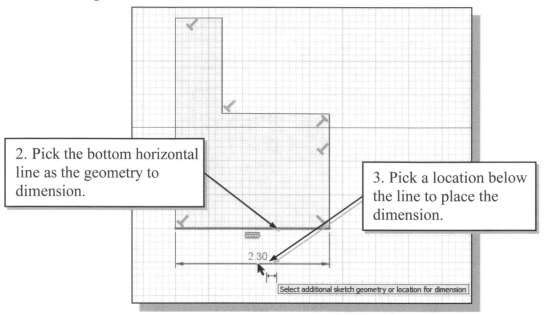

2. Pick the bottom horizontal line as the geometry to dimension.

3. Pick a location below the line to place the dimension.

2.30

Select additional sketch geometry or location for dimension

3. Move the graphics cursor below the selected line and left-click to place the dimension. (Note that the value displayed on your screen might be different than what is shown in the figure above.)

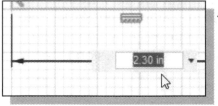

2.30 in

4. Accept the default value by hitting the **Enter** key once.

❖ The **Sketch Dimension** command will create a length dimension if a single line is selected.

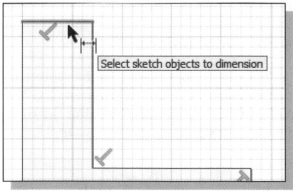

Select sketch objects to dimension

5. Select the top-horizontal line as shown in the figure.

6. Select the bottom-horizontal line as the second entity to dimension as shown.

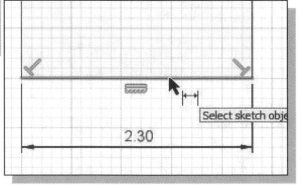

Select sketch obje

2.30

7. Pick a location to the left of the sketch to place the dimension.

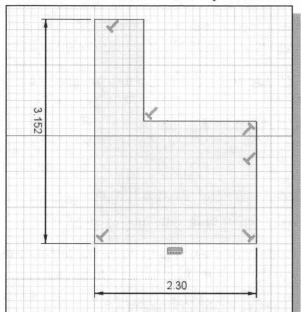

8. Accept the default value by hitting the **Enter** key once.

❖ When two parallel lines are selected, the Sketch Dimension command will create a dimension measuring the distance between them.

9. On your own, repeat the above steps and create additional dimensions (accepting the default values created by Fusion 360) so that the sketch appears as shown. Hit the [ESC] key once to end the Dimension command before proceeding to the next page.

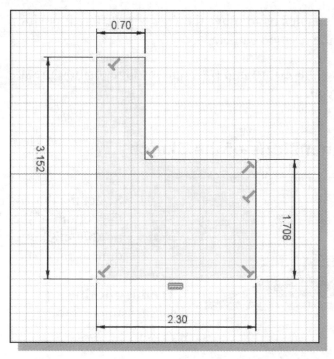

Dynamic Viewing Functions – Zoom and Pan

Autodesk Fusion 360 provides a special user interface called *Dynamic Viewing* that enables convenient viewing of the entities in the graphics window.

1. Click on the **Zoom Window** icon located in the *Navigation* bar as shown.

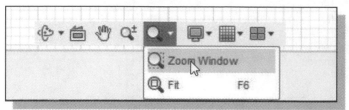

2. Move the cursor near the center of the graphics window.

3. Inside the graphics window, **press and hold down the left-mouse-button**, then move downward to enlarge the current display scale factor.

4. Press the [**Esc**] key once to exit the Zoom command.

5. Click on the **Pan** icon located next to the Zoom command in the *Navigation* bar. The icon is the picture of a hand.

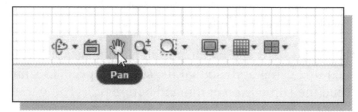

➢   The Pan command enables us to move the view to a different position. This function acts as if you are using a video camera.

6. On your own, use the Zoom and Pan options to reposition the sketch near the center of the screen.

Modifying the Dimensions of the Sketch

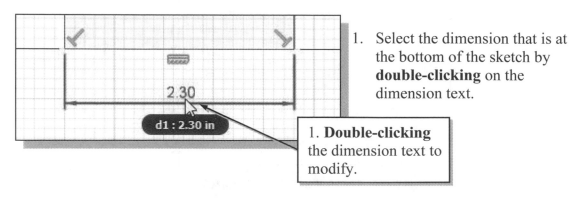

1. Select the dimension that is at the bottom of the sketch by **double-clicking** on the dimension text.

1. **Double-clicking** the dimension text to modify.

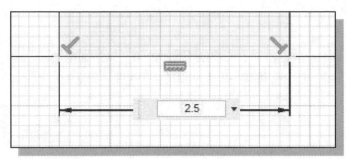

2. In the *Edit Dimension* window, the current length of the line is displayed. Enter **2.5** to set the length of the line.

3. Hit on the **Enter** key once to accept the entered value.

➢ Autodesk Fusion 360 will now update the profile with the new dimension value.

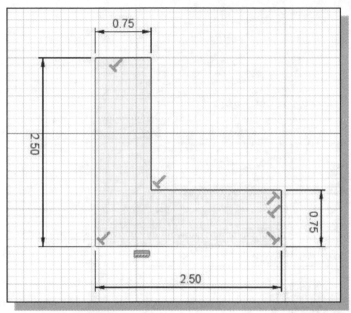

4. On your own, repeat the above steps and adjust the dimensions so that the sketch appears as shown.

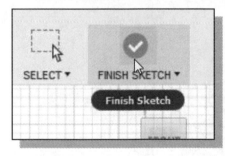

5. In the *Ribbon* toolbar, click once with the **left-mouse-button** on **Finish Sketch** to end the Sketch option.

Step 3: Completing the Base Solid Feature

Now that the 2D sketch is completed, we will proceed to the next step: create a 3D part from the 2D profile. Extruding a 2D profile is one of the common methods that can be used to create 3D parts. We can extrude planar faces along a path. We can also specify a height value and a tapered angle. In Autodesk Fusion 360, each face has a positive side and a negative side; the current face we're working on is set as the default positive side. This positive side identifies the positive extrusion direction and it is referred to as the face's ***normal***.

1. In the *Create toolbar*, select the **Extrude** command by clicking the left-mouse-button on the icon as shown.

2. The 2D sketch we just created is pre-selected as the profile as shown.

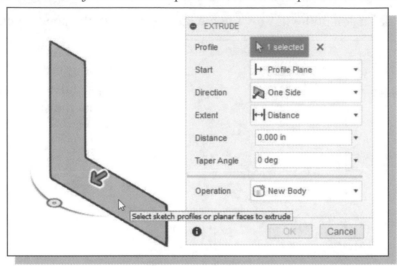

3. In the *Extrude* edit box, enter **2.5** as the extrusion distance.

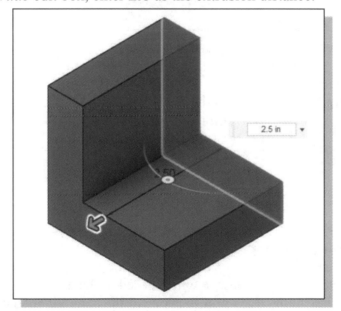

4. Click the **OK** button to proceed with creating the 3D part.

➢ Note that all dimensions disappeared from the screen. All parametric definitions are stored in the database and can be re-displayed.

Dynamic Rotation of the 3D Object – Free Orbit

The **Free Orbit** command allows us to:
- Orbit a part or assembly in the graphics window. Rotation can be around the center mark, free in all directions, or around the X/Y-axes in the *3D-Orbit* display.
- Reposition the part or assembly in the graphics window.
- Display isometric or standard orthographic views of a part or assembly.

The **Free Orbit** tool is accessible while other tools are active. Autodesk Fusion 360 remembers the last used mode when you exit the **Orbit** command.

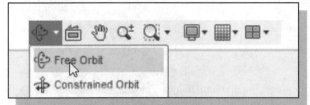

1. Click on the **Free Orbit** icon in the *Navigation* bar.

➤ The *3D Orbit* display is a circular rim with four handles and a center mark. *3D Orbit* enables us to manipulate the view of 3D objects by clicking and dragging with the left-mouse-button:

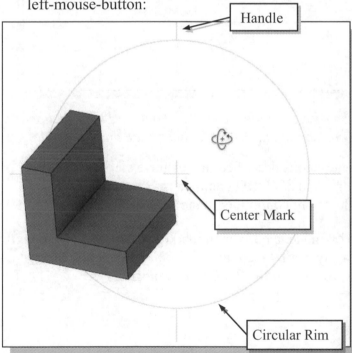

- Drag with the left-mouse-button near the center for free rotation.

- Drag on the handles to orbit around the horizontal or vertical axes.

- Drag on the rim to orbit about an axis that is perpendicular to the displayed view.

- Single left-click to align the center mark of the view.

2. Inside the *circular rim*, press down the left-mouse-button and drag in an arbitrary direction; the **3D Orbit** command allows us to freely orbit the solid model.

3. Move the cursor near the circular rim and notice the cursor symbol changes to a single circle. Drag with the left-mouse-button to orbit about an axis that is perpendicular to the displayed view.

4. Single left-click near the top-handle to align the selected location to the center mark in the graphics window.

5. Activate the **Constrained Orbit** option by clicking on the associated icon as shown.

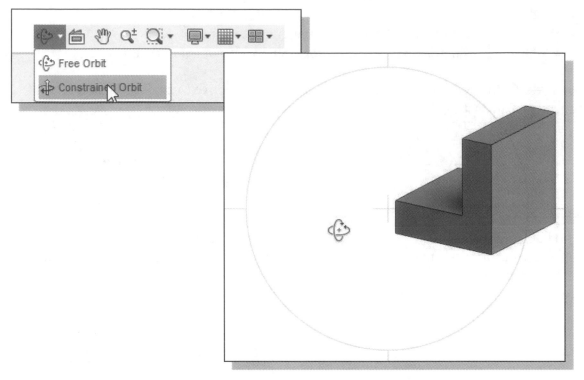

❖ *The Constrained Orbit can be used to rotate the model about axes in Model Space, equivalent to moving the eye position about the model in latitude and longitude.*

6. On your own, use the different options described in the above steps and familiarize yourself with both of the 3D Orbit commands. Reset the display roughly to the *Isometric* angle before continuing to the next section.

❖ Note that while in the 3D Orbit mode, a horizontal marker will be displayed next to the cursor if the cursor is away from the circular rim. This is the exit marker. Left-clicking once will allow you to exit the 3D Orbit command.

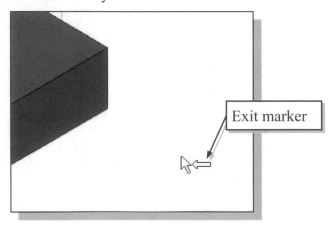

Exit marker

Dynamic Viewing - Quick Keys

We can also use the mouse and the keyboard to access the *Dynamic Viewing* functions. Below are the default Autodesk Fusion 360 quick keys settings; we can use other settings if preferred.

❖ Panning – Press and drag with the mouse wheel

Pressing and dragging with the mouse wheel can also reposition the display. This allows you to reposition the display while maintaining the same scale factor of the display.

Pan   ⇦ | MOUSE WHEEL: PRESS & DRAG | ⇨

❖ Zooming – Turning the mouse wheel

Turning the mouse wheel can also adjust the scale of the display. Turning forward will reduce the scale of the display, making the entities display smaller on the screen. Turning backward will magnify the scale of the display.

Zoom | MOUSE WHEEL: TURNING |

❖ 3D Rotation – Shift and the middle-mouse-button

Hold the **Shift** key down and drag with the middle-mouse-button to orbit the display.

3D Rotation | Shift | + ⇦ | MOUSE | ⇨

Viewing Tools – Display Control bar and View Cube

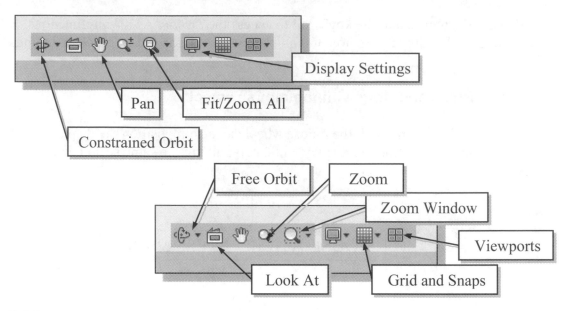

Orbit – In a part or assembly, adds an orbit symbol and cursor to the view. You can orbit the view planar to the screen around the center mark, around a horizontal or vertical axis, or around the X and Y axes.

Look At – In a part or assembly, zooms and orbits the model to display the selected element planar to the screen or a selected edge or line horizontal to the screen. (Not used in drawings.)

Fit/Zoom All – Adjusts the view so that all items on the screen fit inside the graphics window.

Zoom Window – Use the cursor to define a region for the view; the defined region is zoomed to fill the graphics window.

Zoom – Moving upward will reduce the scale of the display, making the entities display smaller on the screen. Moving downward will magnify the scale of the display.

Pan – This allows you to reposition the display while maintaining the same scale factor of the display.

Display settings – This panel contains multiple control to the display related settings, such as Visual Style and Camera.

View Cube

View Cube – The View Cube is a 3D navigation tool that appears at the upper right corner of the *Fusion 360* main window. The View Cube is a clickable interface which allows you to switch between standard and isometric views.

The View Cube also provides visual feedback about the current viewpoint of the model as view changes occur. When the cursor is positioned over the View Cube, it becomes active and allows you to switch to one of the available preset views, roll the current view, or change to the Home view of the model.

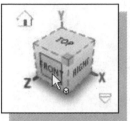

1. Move the cursor over the ViewCube and notice the different sides of the ViewCube become highlighted and can be activated.

2. Single left-click when the front side is activated as shown. The current view is set to view the front side.

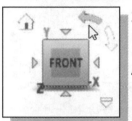

3. Move the cursor over the counter-clockwise arrow of the ViewCube and notice the orbit option becomes highlighted.

4. Single left-click to activate the counter-clockwise option as shown. The current view is orbited 90 degrees; we are still viewing the front side.

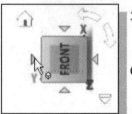

5. Move the cursor over the left arrow of the ViewCube and notice the orbit option becomes highlighted.

6. Single left-click to activate the left arrow option as shown. The current view is now set to view the top side.

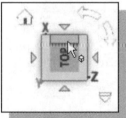

7. Move the cursor over the top edge of the ViewCube and notice the roll option becomes highlighted.

8. Single left-click to activate the roll option as shown. The view will be adjusted to roll 45 degrees.

9. Move the cursor over one of the corners of the ViewCube and drag with the left-mouse-button to activate the **Free Rotation** option.

10. Move the cursor over the home icon of the ViewCube and notice the Home View option becomes highlighted.

11. Single left-click to activate the **Home View** option as shown. The view will be adjusted back to the default *isometric view*.

Display Modes

The **Visual Style** in the *Display Settings* list has six display-modes showing shaded renderings and wireframe representations of the model.

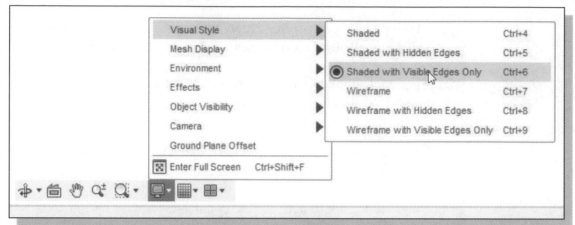

❖ Shaded Solid Modes:

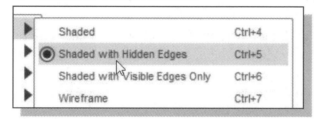

The *Shaded Solid* display modes generate high quality shaded images of the 3D object.

❖ Wireframe Display Modes:

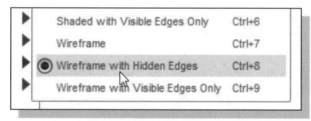

The *Wireframe Image* display options allow the display of the 3D objects using the wireframe representation scheme.

Orthographic vs. Perspective

Besides the above basic display modes, we can also choose orthographic view using the parallel edges representation scheme or perspective view using the perspective, nonparallel edges, and representation scheme of the display. Click on the **Camera** list in the *Display Settings* to select the different display options as shown in the figure.

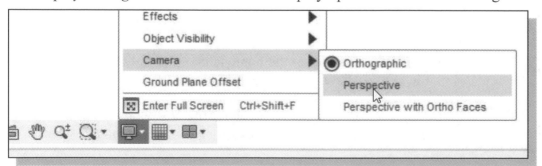

Step 4-1: Adding an Extruded Feature

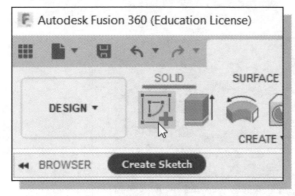

1. In the *Create* toolbar, select the **Create Sketch** command by left-clicking once on the icon.

2. Next to the cursor, the message "*Select a plane or planar face*" is displayed. Autodesk Fusion 360 expects us to identify a planar surface where the 2D sketch of the next feature is to be created. Move the graphics cursor on the 3D part and notice that Autodesk Fusion 360 will automatically highlight feasible planes and surfaces as the cursor is on top of the different surfaces. Pick the top horizontal face of the 3D solid object.

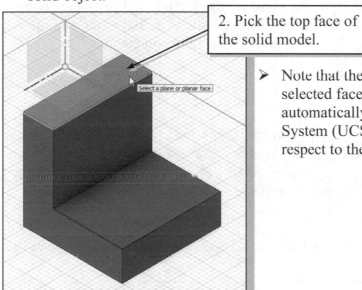

2. Pick the top face of the solid model.

➤ Note that the sketch plane is aligned to the selected face. Autodesk Fusion 360 automatically establishes a User-Coordinate-System (UCS) and records its location with respect to the part on which it was created.

• Next, we will create and profile another sketch, a rectangle, which will be used to create another extrusion feature that will be added to the existing solid object.

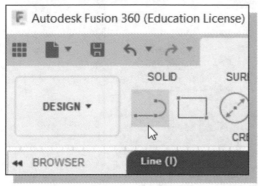

3. Select the **Line** command by clicking once with the **left-mouse-button** on the icon in the *Sketch* tab on the Ribbon.

4. Create a sketch, starting at the top left corner of the solid model, with six line segments perpendicular/parallel to the solid model as shown below.

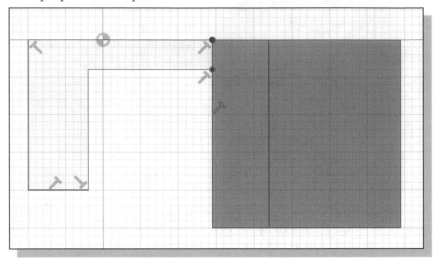

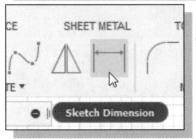

5. Inside the graphics window, click once with the **right-mouse-button** to display the option menu. Select **Cancel (ESC)** in the pop-up menu to end the Line command.

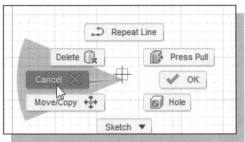

6. Select the **Sketch Dimension** command in the *Sketch* toolbar. The Sketch Dimension command allows us to quickly create and modify dimensions. You can also hit [D] on the keyboard to activate this command.

7. Create the four size dimensions, set values to **2.5 & .75**, to describe the size of the sketch as shown.

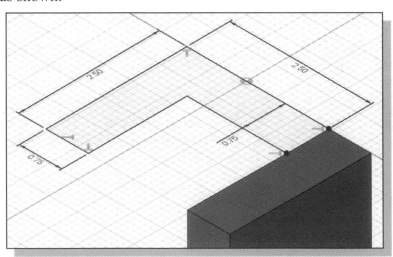

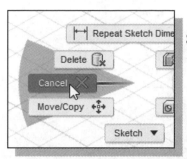

8. Inside the graphics window, click once with the **right-mouse-button** to display the option menu. Select **Cancel** in the pop-up menu to end the Sketch Dimension command.

9. In the *Ribbon* area, select **Finish Sketch** to end the Sketch command.

10. In the *Create toolbar,* select the **Extrude** command by left-clicking on the icon.

11. Select inside the 2D sketch we just created as the region to extrude.

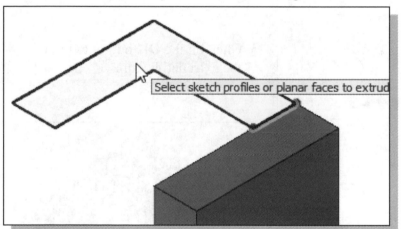

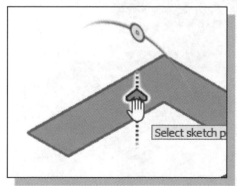

12. Move the cursor on top of the arrow; notice the dashed line appears as shown. We can drag the arrow, with the left mouse button, to set the extrusion distance and direction.

13. Drag the arrow **downward**, with the left mouse button, and set the extrusion distance to **-2.5** as shown.

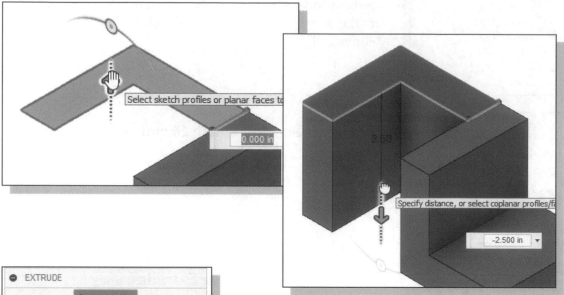

14. In the *Extrude* pop-up window, confirm the settings are set to **Join** and **-2.50** in as shown.

15. Click on the **OK** button to proceed with creating the extruded feature.

Step 4-2: Adding a Cut Feature

Next, we will create and profile a circle, which will be used to create a cut feature that will be added to the existing solid object.

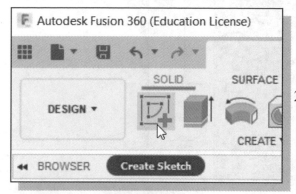

1. In the *Create* toolbar, select the **Create Sketch** command by left-clicking once on the icon.

2. Click on the **Minimize** icon to minimize the *Sketch Palette*.

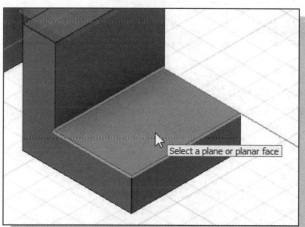

3. Pick the top horizontal face of the 3D solid model as shown.

➤ Note that the sketch plane is aligned to the selected face. Autodesk Fusion 360 automatically establishes a User-Coordinate-System (UCS) and records its location with respect to the part on which it was created.

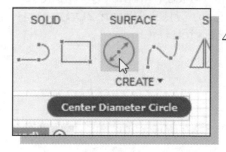

4. Select the **Center Diameter circle** command by clicking once with the **left-mouse-button** on the icon in the *Sketch* toolbar as shown.

5. Create a circle of arbitrary size on the top face of the solid model as shown.

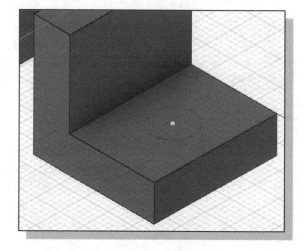

6. On your own, create and modify the dimensions of the sketch as shown in the figure.

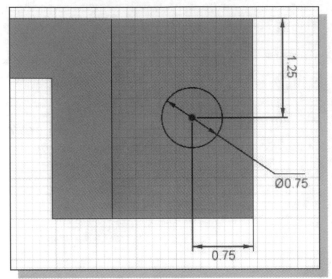

7. Inside the graphics window, click once with the **right-mouse-button** to display the option menu. Select **Cancel** in the pop-up menu to end the Sketch Dimension command.

8. Inside the graphics window, click once with the **right-mouse-button** to display the option menu. Select **Finish 2D Sketch** in the pop-up menu to end the Sketch option.

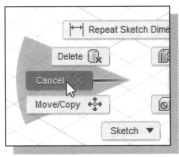

9. In the *Create toolbar,* select the **Extrude** command by left-clicking on the icon.

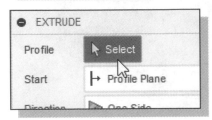

10. In the *Extrude* pop-up control, the **Profile** button is activated; Autodesk Fusion 360 expects us to identify the profile to be extruded.

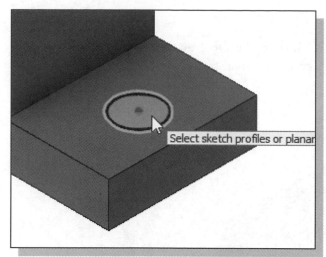

11. Click on the inside of the sketched circle to define the extrude profile as shown.

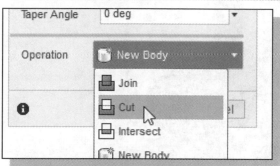

12. Select the **CUT** option in the *feature option* list to set the extrusion operation to *Cut*.

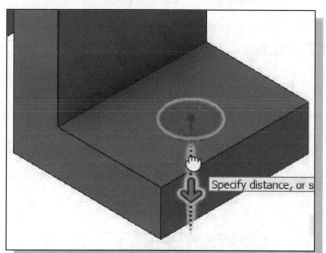

13. Move the cursor on top of the arrow; notice the dashed line appears.

14. Drag the arrow **downward**, with the left mouse button, to set the extrusion direction.

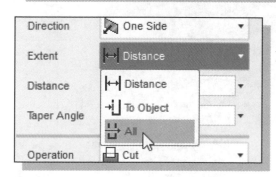

15. Set the *Extent* option to **Through All** as shown. The *All* option instructs the software to calculate the extrusion distance and assures the created feature will always cut through the full length of the model.

16. Click on the **OK** button to proceed with creating the extruded feature.

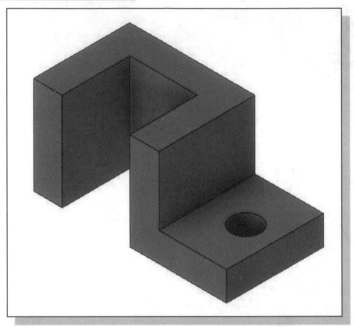

Step 4-3: Adding another Cut Feature

Next, we will create and profile a triangle, which will be used to create a cut feature that will be added to the existing solid object.

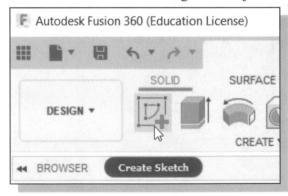

1. In the *3D Model* tab select the **Create Sketch** command by left-clicking once on the icon.

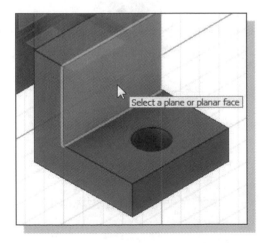

2. Pick the vertical face of the 3D solid model next to the horizontal section as shown.

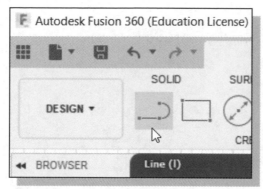

3. Select the **Line** command by clicking once with the **left-mouse-button** on the icon in the *Sketch* tab.

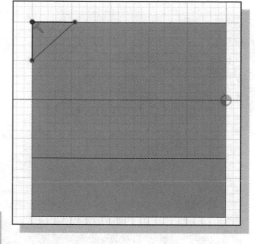

4. Start at the upper left corner and create **three line segments** to form a small triangle as shown.

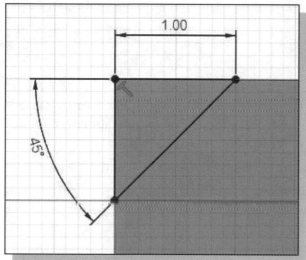

5. On your own, create and modify the two dimensions of the sketch as shown in the figure. (Hint: create the angle dimension by selecting the two adjacent lines and place the angular dimension inside the desired quadrant.)

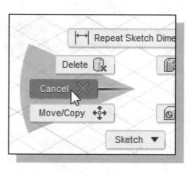

6. Inside the graphics window, click once with the **right-mouse-button** to display the option menu. Select **Cancel** in the pop-up menu to end the Sketch Dimension command.

7. In the *ribbon toolbar area, s*elect **Stop Sketch** to end the Sketch option.

8. In the *Ribbon* toolbar, select the **Extrude** command by left-clicking the icon.

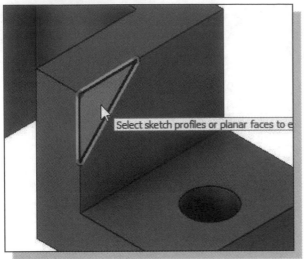

9. Click on the inside of the sketched circle to define the extrude profile as shown.

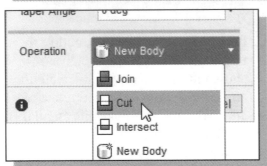

10. Select the **CUT** option in the *feature option list* to set the extrusion operation to *Cut*.

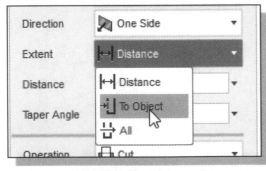

11. Set the *Extents* option to **To Object** as shown. The *To Object* option instructs the software to calculate the extrusion distance and assures the created feature will always cut through the proper length of the model.

• Note the Select icon is activated and Fusion 360 expects us to select an object to set the termination of the extrude feature.

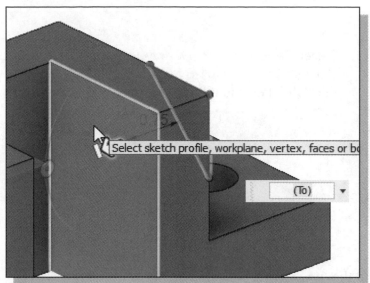

12. On your own, use the dynamic rotate function and select the back surface as shown.

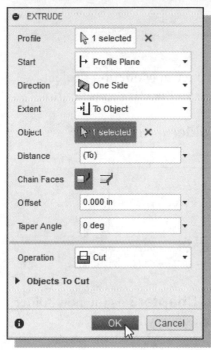

13. On your own, confirm the settings in the extrude dialog box are as shown and click on the **OK** button to proceed with creating the extruded feature.

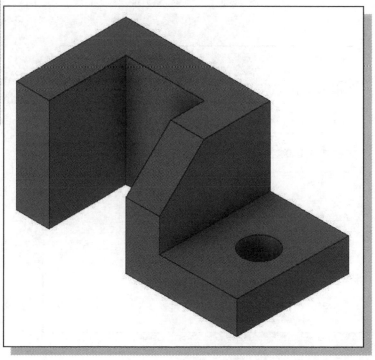

Save the Model

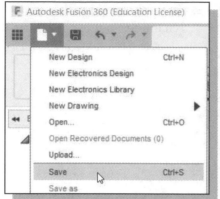

1. Select **Save** in the *Quick Access* toolbar, or you can also use the "**Ctrl-S**" combination (hold down the "Ctrl" key and hit the "S" key once) to save the part.

2. Expand the *option* list by clicking on the *down arrow* as shown.

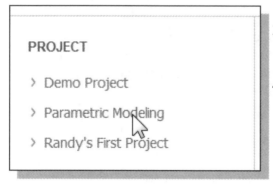

3. Select the Parametric Modeling Project by clicking once with the left mouse button.

4. Click **New Folder** to create a new subfolder.

5. Enter **Chapter2** as the new folder name as shown.

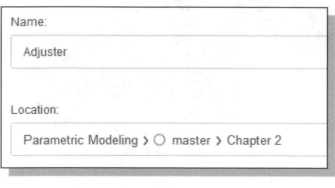

6. **Double-click** on the Chapter2 folder to open it.

7. In the *file name* editor box, enter **Adjuster** as the file name.

8. Click on the **Save** button to save the file.

Review Questions:

1. What is the first thing we should set up in Autodesk Fusion 360 when creating a new model?

2. Describe the general *parametric modeling* procedure.

3. Describe the general guidelines in creating *Rough Sketches*.

4. List two of the geometric constraint symbols used by Autodesk Fusion 360.

5. What was the first feature we created in this lesson?

6. How many solid features were created in the tutorial?

7. How do we control the size of a feature in parametric modeling?

8. Which command was used to create the last cut feature in the tutorial? How many dimensions do we need to fully describe the cut feature?

9. List and describe three differences between parametric modeling and traditional 2D Computer Aided Drafting techniques.

Exercises: Create and save the exercises in the Chapter2 folder.
(All dimensions are in inches.)

1. **Inclined Support** (Thickness: **.5**)

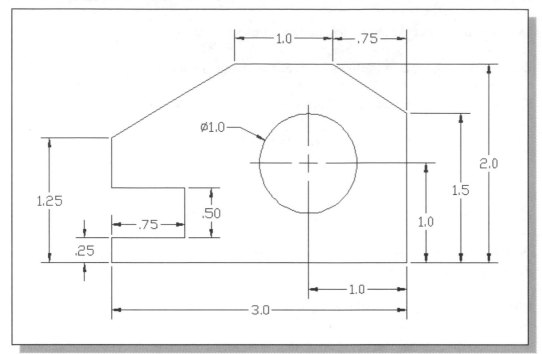

2. **Spacer Plate** (Thickness: **.125**)

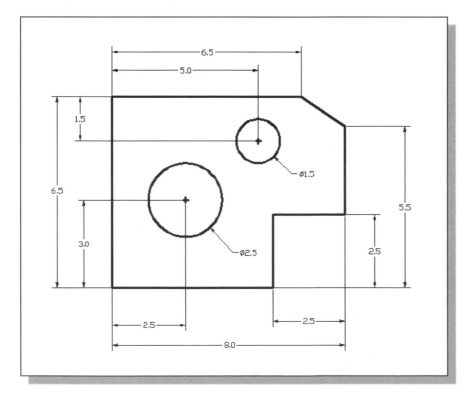

3. **Positioning Stop**

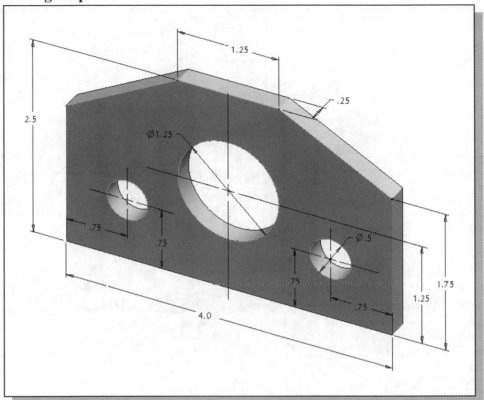

4. **Guide Block**

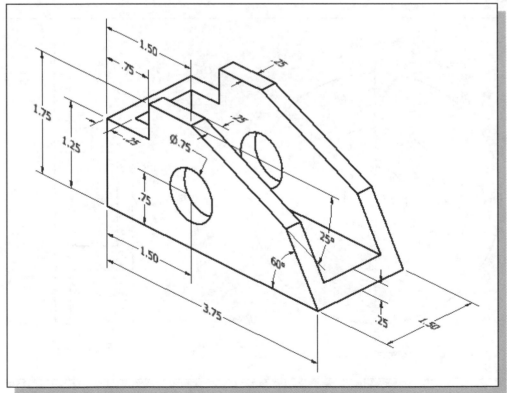

5. **Slider Block**

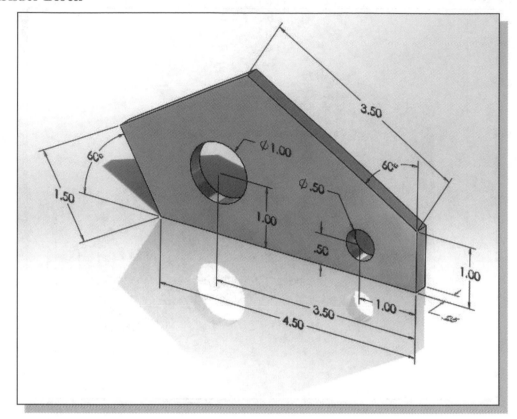

6. **Coupler**

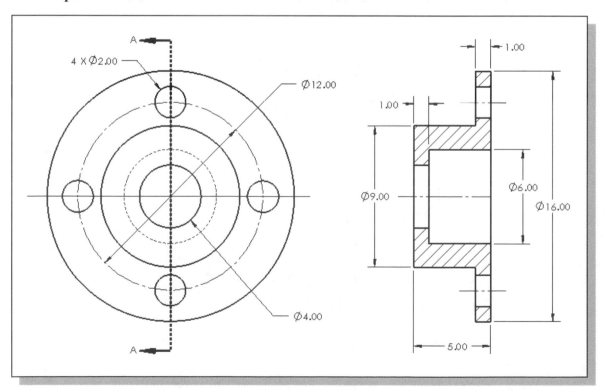

Chapter 3
Constructive Solid Geometry Concepts

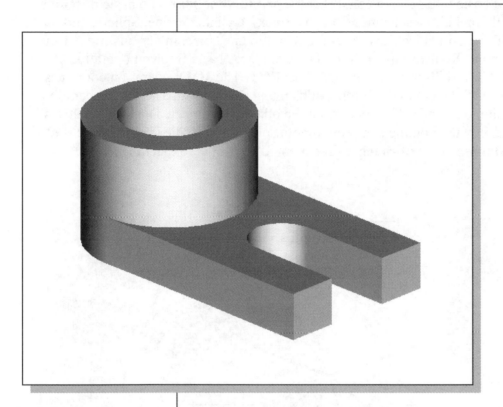

Learning Objectives

- ♦ **Understand Constructive Solid Geometry Concepts**
- ♦ **Create a Binary Tree**
- ♦ **Understand the Basic Boolean Operations**
- ♦ **Set up GRID and SNAP Intervals**
- ♦ **Understand the Importance of Order of Features**
- ♦ **Create Placed Features**
- ♦ **Use the Different Extrusion Options**

Introduction

In the 1980s, one of the main advancements in **solid modeling** was the development of the **Constructive Solid Geometry** (CSG) method. CSG describes the solid model as combinations of basic three-dimensional shapes (**primitive solids**). The basic primitive solid set typically includes Rectangular-prism (Block), Cylinder, Cone, Sphere, and Torus (Tube). Two solid objects can be combined into one object in various ways using operations known as **Boolean operations**. There are three basic Boolean operations: **JOIN (Union)**, **CUT (Difference)**, and **INTERSECT**. The **JOIN** operation combines the two volumes included in the different solids into a single solid. The **CUT** operation subtracts the volume of one solid object from the other solid object. The **INTERSECT** operation keeps only the volume common to both solid objects. The CSG method is also known as the **Machinist's Approach**, as the method is parallel to machine shop practices.

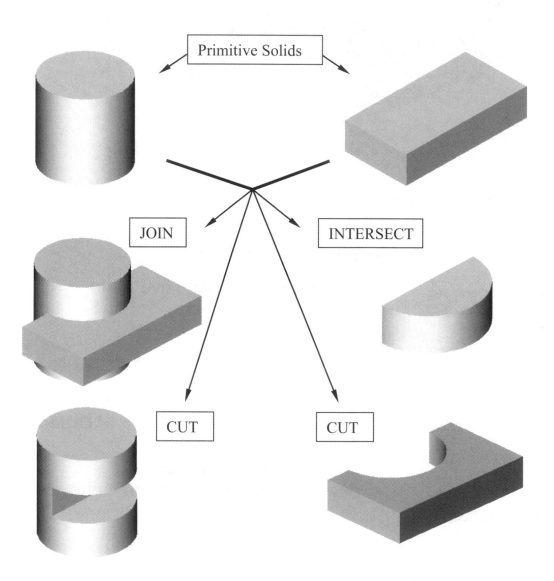

Binary Tree

The CSG is also referred to as the method used to store a solid model in the database. The resulting solid can be easily represented by what is called a **binary tree**. In a binary tree, the terminal branches (leaves) are the various primitives that are linked together to make the final solid object (the root). The binary tree is an effective way to keep track of the *history* of the resulting solid. By keeping track of the history, the solid model can be re-built by re-linking through the binary tree. This provides a convenient way to modify the model. We can make modifications at the appropriate links in the binary tree and re-link the rest of the history tree without building a new model.

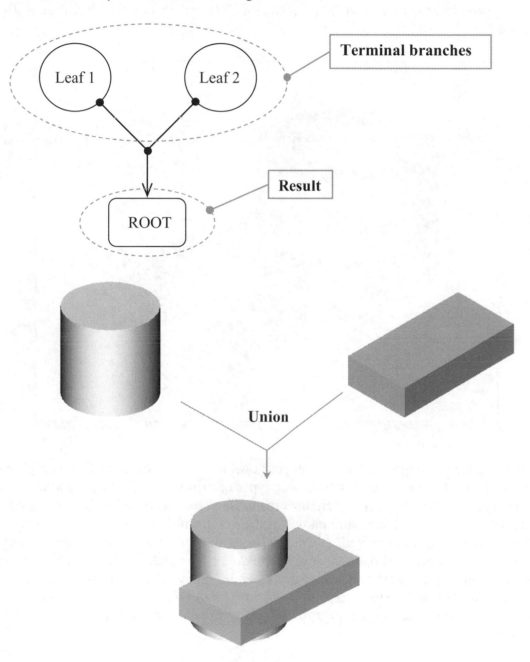

The Locator Design

The CSG concept is one of the important building blocks for feature-based modeling. In Autodesk Fusion 360, the CSG concept can be used as a planning tool to determine the number of features that are needed to construct the model. It is also a good practice to create features that are parallel to the manufacturing process required for the design. With parametric modeling, we are no longer limited to using only the predefined basic solid shapes. In fact, any solid features we create in Autodesk Fusion 360 are used as primitive solids; parametric modeling allows us to maintain full control of the design variables that are used to describe the features. In this lesson, a more in-depth look at the parametric modeling procedure is presented. The equivalent CSG operation for each feature is also illustrated.

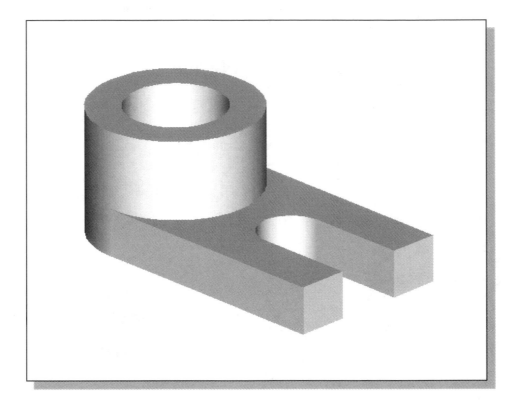

> ➢ Before going through the tutorial, on your own make a sketch of a CSG binary tree of the *Locator* design using only two basic types of primitive solids: cylinder and rectangular prism. In your sketch, how many *Boolean operations* will be required to create the model? What is your choice of the first primitive solid to use, and why? Take a few minutes to consider these questions and do the preliminary planning by sketching on a piece of paper. Compare the sketch you make to the CSG binary tree steps shown on the next page. Note that there are many different possibilities in combining the basic primitive solids to form the solid model. Even for the simplest design, it is possible to take several different approaches to creating the same solid model.

Modeling Strategy – CSG Binary Tree

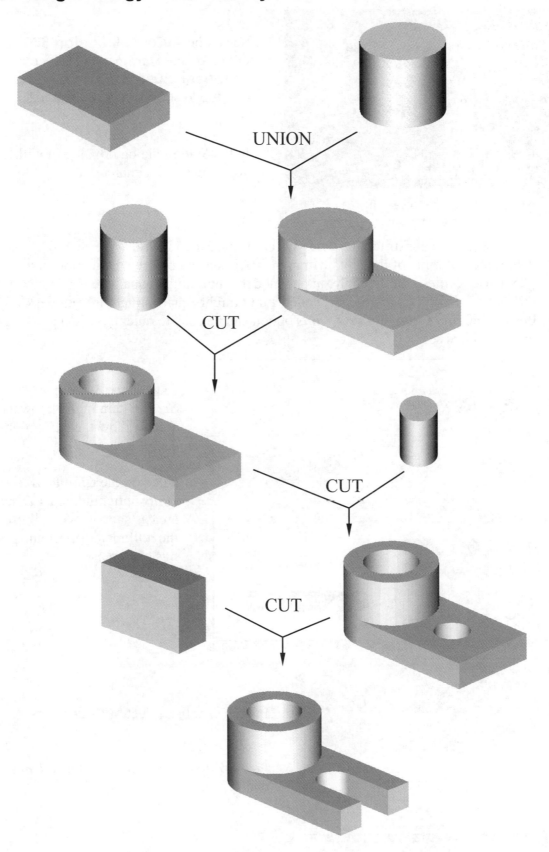

Starting Autodesk Fusion 360

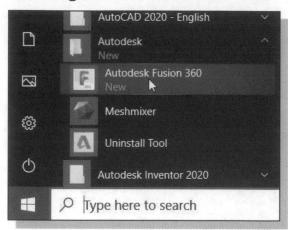

1. Select the **Autodesk Fusion 360** option on the *Start* menu or select the **Autodesk Fusion 360** icon on the desktop to start Autodesk Fusion 360.

2. In the *Sign In* dialog box, log in with your email or username.

❖ Every object we construct in a CAD system is measured in units. We should determine the value of the units within the CAD system before creating the first geometric entities. For example, in one model, a unit might equal one millimeter of the real-world object; in another model, a unit might equal an inch. In Autodesk Fusion 360, the *Change Active Units* option can be used to quickly switch to using a different units set.

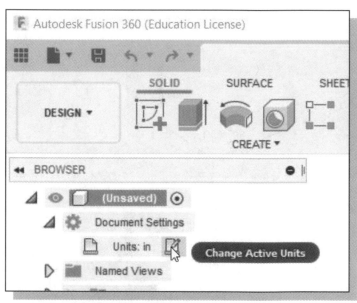

3. Expand the *Document Setting* item in the Browser area to show the current units setup as shown.

4. Click on the *Change Active Units* option in the Browser area as shown. We will use the millimeter (mm) units for this example.

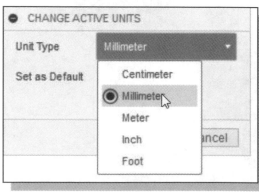

5. In the *Units* list, select **Millimeter** as shown.

6. Pick **OK** to exit the *Change Active Units* dialog box.

Model Dimensions Format

- Autodesk Fusion 360 also provides controls to the display of the model dimensions based on the screen or in a drawing.

1. In the *Account name* area, select **[Preferences]**.

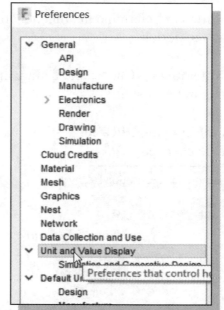

2. In the *Preferences* dialog box, choose the **Unit and Value Display** option as shown in the figure.

3. Set the precision to ***no digits*** after the decimal point for both the *linear dimension* and *angular dimension* displays as shown in the figure.

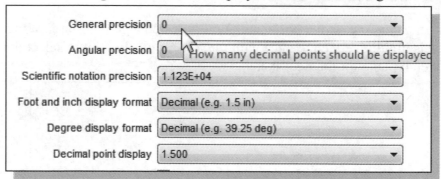

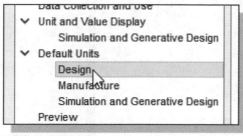

4. Note that the **Default Units** for different applications can also be set under the *Default Units* list.

5. Pick **Apply** and then pick **OK** to exit the *Preferences Settings* dialog box.

Base Feature

In *parametric modeling*, the first solid feature is called the **base feature**, which usually is the primary shape of the model. Depending upon the design intent, additional features are added to the base feature.

These are some of the considerations involved in selecting the base feature:

- **Design intent** – Determine the functionality of the design; identify the feature that is central to the design.

- **Order of features** – Choose the feature that is the logical base in terms of the order of features in the design.

- **Ease of making modifications** – Select a base feature that is more stable and is less likely to be changed.

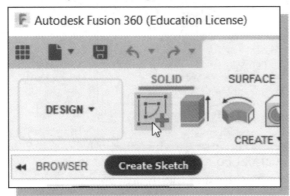

1. Activate the **Create Sketch** icon with a single click of the left-mouse-button.

2. Move the cursor over the horizontal *XZ Plane* in the graphics area. When the *XZ Plane* is highlighted, click once with the **left-mouse-button** to select the *Plane* as the sketch plane for the new sketch.

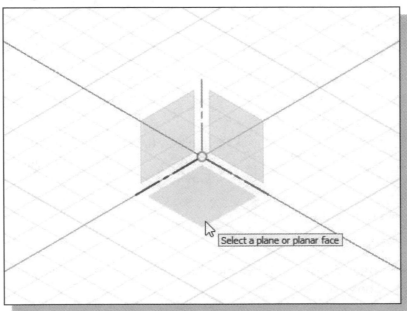

➢ A rectangular block will be first created as the base feature of the *Locator* design.

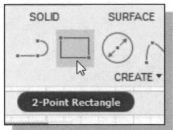

3. Select the **2-point rectangle** command by clicking on the icon once with the **left-mouse-button**.

4. Create a rectangle of arbitrary size by selecting two locations on the screen as shown below.

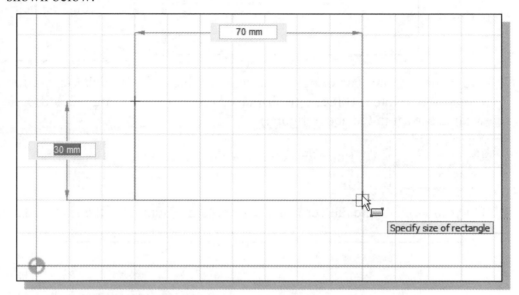

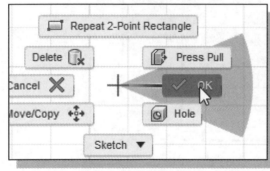

5. Inside the graphics window, click once with the **right-mouse-button** to bring up the option menu.

6. Select **OK** to cnd the Rectangle command.

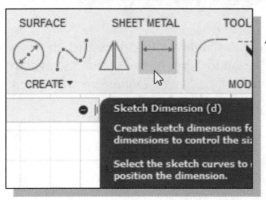

7. In the sketch toolbar, select the **Sketch Dimension** command as shown. (You can also hit the [**D**] key.)

8. Select the bottom horizontal line by left-clicking once on the line.

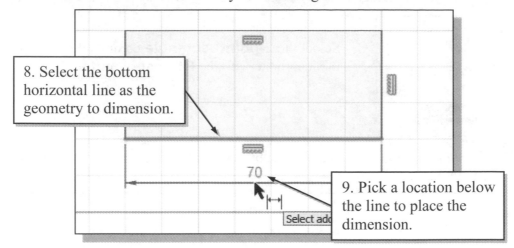

8. Select the bottom horizontal line as the geometry to dimension.

70

9. Pick a location below the line to place the dimension.

9. Move the graphics cursor below the selected line and left-click to place the dimension. (Note that the value displayed on your screen might be different than what is shown in the above figure.)

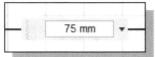

10. Enter **75** as the new dimension value as shown.

11. On your own, create the vertical size dimension (30mm) of the sketched rectangle as shown.

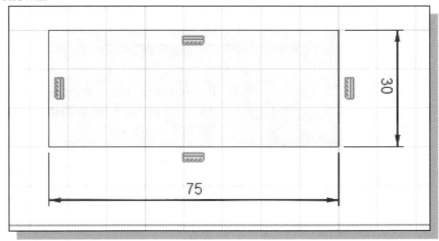

30

75

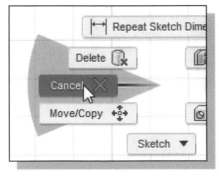

12. Inside the graphics window, click once with the right-mouse-button to bring up the option menu and click **Cancel** to end the *Sketch Dimension* command.

Modify existing Dimensions

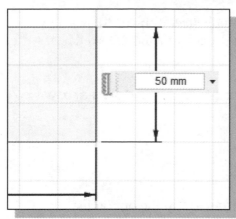

1. Select the height dimension that is to the right side of the sketch by ***double-clicking*** with the left-mouse-button on the dimension text.

2. In the *Edit Dimension* window, the current length of the line is displayed. Enter **50** to set the selected length of the sketch to 50 millimeters.

3. Hit the **Enter key** to accept the entered value.

➢ Autodesk Fusion 360 will now update the profile with the new dimension value.

4. On your own, experiment with adjusting the dimensions; set the dimensions as shown below before proceeding to the next step.

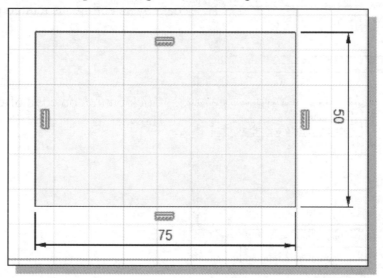

Reposition Dimensions

1. Move the cursor near the vertical dimension; note that the dimension is highlighted.

2. Drag with the left-mouse-button to reposition the selected dimension.

3. Repeat the above steps to reposition the horizontal dimension.

Using the Measure Tool

Autodesk Fusion 360 also provides several measuring tools that allow us to measure distance, perimeter, area and additional information of the constructed 2D sketches.

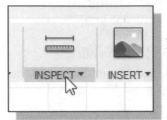

1. In the *Inspect icon*, left-click once on the **down arrow** to open up the option list.

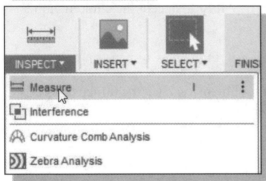

2. Select **Measure** in the option list as shown.

    •   Note that we can use the measure command to measure distance, angle, area or position data by selecting associated objects.

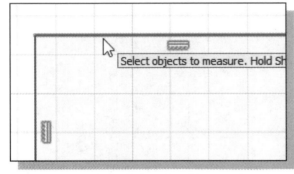

3. Click on the top edge of the rectangle as shown.

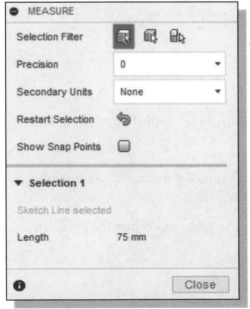

4. The associated length measurement of the selected geometry, **75 mm**, is displayed in the *Measure* dialog box as shown.

5. Click outside the rectangle, or hold down the **[Ctrl]** key and click on the top edge of the rectangle, to *de-select* the object.

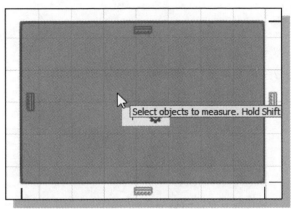

6. Click on the inside region of the rectangle as shown.

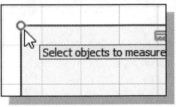

7. The associated area and loop length of the selected region is calculated and displayed in the *Measure* dialog box as shown.

8. On your own, **de-select** the region.

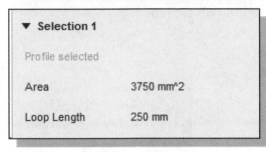

9. Click on the upper left corner of the rectangle as shown.

10. Autodesk Fusion 360 now displays the location information of the selected corner.

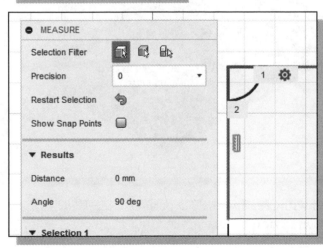

11. On your own, use the **Measure** command and measure one of the inside corners of the region.

12. Click **Close** to exit the Measure command.

Completing the Base Solid Feature

1. In the *Ribbon* toolbar, click once with the **left-mouse-button** on **Finish Sketch** to end the Sketch option.

2. In the *Create toolbar* select the **Extrude** command by clicking the left-mouse-button on the icon.

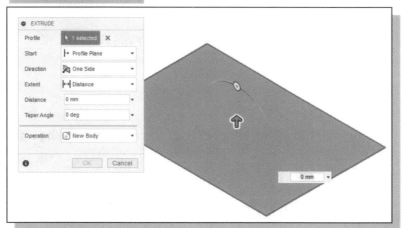

3. Note that the 2D sketch we just created is pre-selected as the region to be extruded.

4. In the *Extrude* pop-up window, enter **15** as the extrusion distance. Notice that the **New Body** option is automatically set for the first solid feature of the design.

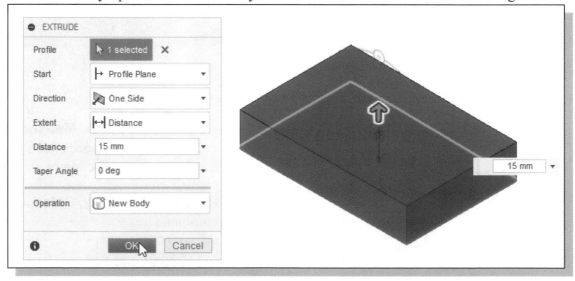

5. Click on the **OK** button to proceed with creating the 3D part. Use the *View Cube* and change the display to the isometric view before proceeding.

Creating the Next Solid Feature

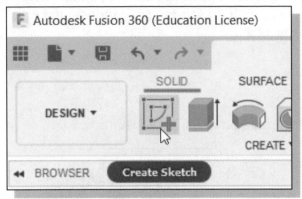

1. In the *Sketch toolbar* select the **Create Sketch** command by left-clicking once on the icon.

2. Autodesk Fusion 360 expects us to identify a planar surface where the 2D sketch of the next feature is to be created. Move the graphics cursor on the 3D part and notice that Autodesk Fusion 360 will automatically highlight feasible planes and surfaces as the cursor is on top of the different surfaces.

3. Use the **View Cube** to adjust the display viewing the bottom face of the solid model as shown below.

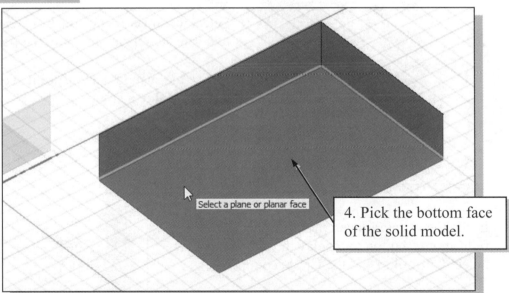

Select a plane or planar face

4. Pick the bottom face of the solid model.

4. Pick the bottom face of the 3D model as the sketching plane.

➢ Note that the sketching plane is aligned to the selected face. Autodesk Fusion 360 automatically establishes a User-Coordinate-System (UCS) and records its location with respect to the part on which it was created.

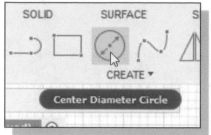

5. Select the **Center Diameter Circle** command by clicking once with the left-mouse-button on the icon in the *Sketch* tab.

➢ We will align the center of the circle to the midpoint of the base feature.

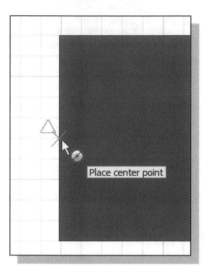

6. Select the midpoint of the left edge when the midpoint constraint symbol is displayed as shown.

• Notice the snap to the **Midpoint** *symbol* is a **Triangle.**

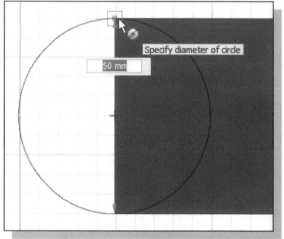

7. Select the top corner of the base feature to create a circle as shown in the figure.

• Notice the snap to the **endpoint** *symbol* is a **circle inside a square.**

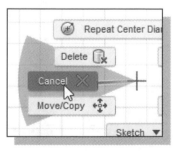

8. Inside the *graphics window*, click once with the right-mouse-button to display the option menu. Select **Cancel** to end the Circle command.

9. In the *Ribbon* toolbar, select **Finish Sketch** to exit the Sketch mode.

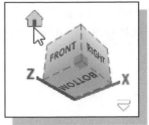

10. Select **Home View** in the **ViewCube** to change the display to the isometric view as shown.

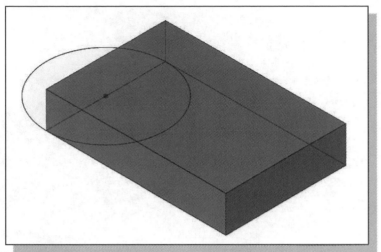

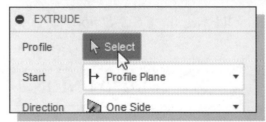

11. In the *Create toolbar*, select the **Extrude** command by left-clicking the icon.

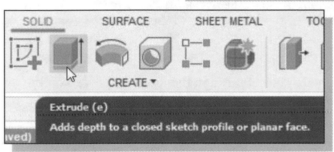

12. In the *Extrude* pop-up control, the **Profile** button is activated as shown; Autodesk Fusion 360 expects us to identify the profile to be extruded. Select the **two semi-circle regions** to define the profile to be extruded.

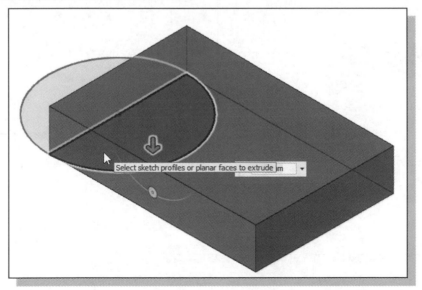

13. In the *Extrude* pop-up control, set the direction of extrusion to upward, and the extrusion distance to **-40** and also set the solid operation to **Join** as shown below.

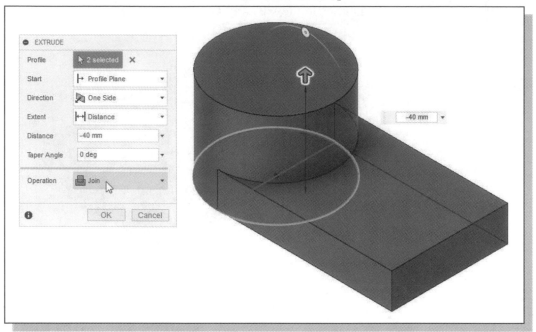

➢ Note that most of the settings can also be set through the icons displayed on the screen.

14. Click on the **OK** button to proceed with the *Join* operation.

● The two features are joined together into one solid part; the *CSG-Union* operation was performed.

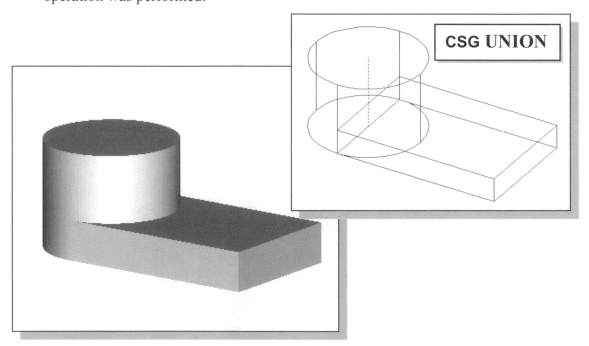

Creating a CUT Feature

- We will create a circular cut as the next solid feature of the design. We will align the sketch plane to the top of the last cylinder feature.

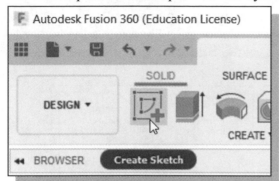

1. In the *Create toolbar* select the **Create Sketch** command by left-clicking once on the icon.

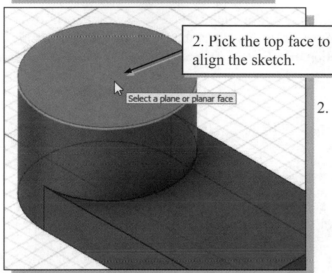

2. Pick the top face to align the sketch.

Select a plane or planar face

2. Pick the top face of the cylinder as shown.

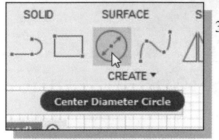

3. Select the **Center Diameter circle** command by clicking once with the **left-mouse-button** on the icon in the *Sketch tab* of the Ribbon toolbar.

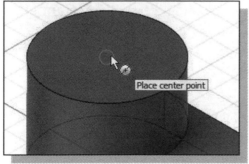

Place center point

4. Select the **Center** point of the top face of the 3D model as shown.

- Notice the snap to the **center** *symbol* is a **circle** as shown.

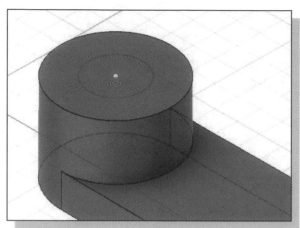

5. Sketch a circle of arbitrary size inside the top face of the cylinder as shown to the left.

6. Use the right-mouse-button to display the option menu and move the cursor on the **Sketch** button to show more options.

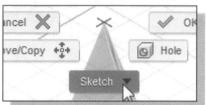

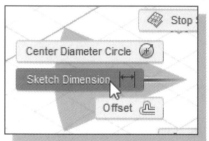

7. Inside the option menu, click once with the **right-mouse-button** to select the **Sketch Dimension** option.

8. Create a dimension to describe the size of the circle and set it to **30mm**.

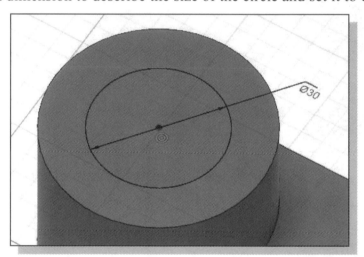

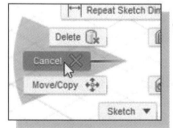

9. Inside the graphics window, click once with the right-mouse-button to display the option menu. Select **Cancel** in the pop-up menu to end the Dimension command.

10. Inside the graphics window, click once with the right-mouse-button to display the option menu. Select **Finish Sketch** in the pop-up menu to end the Sketch option.

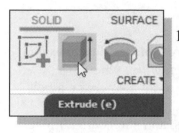

11. In the *Create toolbar*, select the **Extrude** command by left-clicking on the icon.

12. In the *Extrude* pop-up window, the **Profile** button is pressed down; Autodesk Fusion 360 expects us to identify the profile to be extruded. Select **inside** the **circle** we just created as the profile to be extruded.

13. In the *Extrusion* pop-up window, set the operation option to **Cut**. Select **Through All** as the *Extents* option, as shown below.

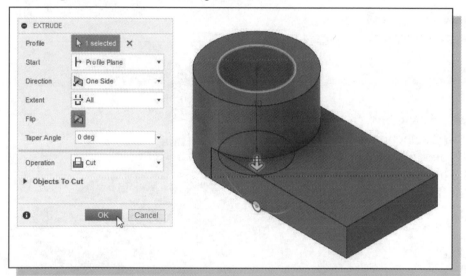

14. Click on the **OK** button to proceed with the *Cut* operation.

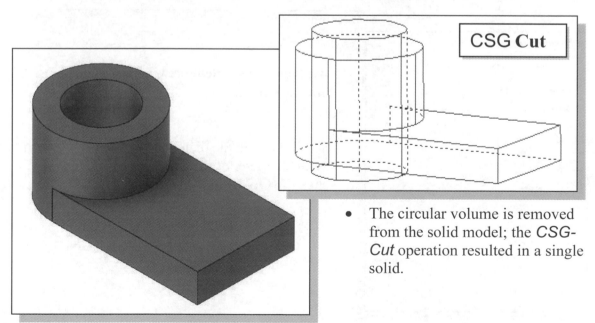

- The circular volume is removed from the solid model; the *CSG-Cut* operation resulted in a single solid.

Creating a Placed Feature

In Autodesk Fusion 360, there are two types of geometric features: **placed features** and **sketched features**. The last cut feature we created is a *sketched feature*, where we created a rough sketch and performed an extrusion operation. We can also create a hole feature, which is a placed feature. A *placed feature* is a feature that does not need a sketch and can be created automatically. Holes, fillets, chamfers, and shells are all placed features.

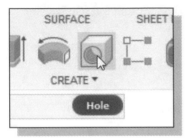

1. In the *Create* toolbar, select the **Hole** command by left-clicking on the icon.

2. In the *Hole* dialog box, the **Face/Point** selection is activated, pick a location inside the top horizontal surface of the base feature as shown.

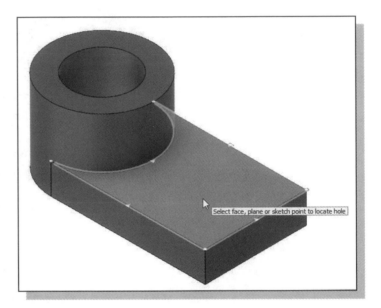

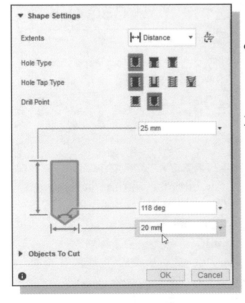

*   By default, a **single hole** will be created using the **Simple** option.

3. Scroll down in the dialog box and enter **20 mm** as the diameter of the hole as shown. **Do Not** click the **OK** button yet.

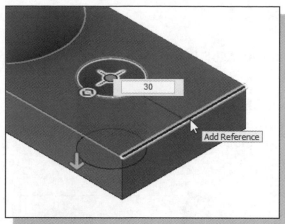

4. Pick the **right-edge** of the top face of the base feature as shown. This will be used as the first reference for placing the hole on the plane.

5. Enter **30** mm as the first reference distance. **Do Not** click the OK button yet.

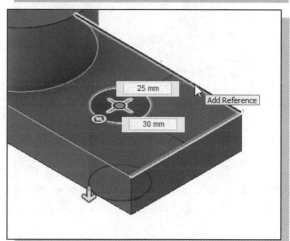

6. Pick the **adjacent top-edge** of the top face as shown. This will be used as the second reference for placing the hole on the plane.

7. Enter **25** mm as the distance as shown.

8. In *Holes* dialog box, set the *Extents* option to **Through All**.

9. Click on the **OK** button to proceed with the *Hole* feature.

• The circular volume is removed from the solid model; the *CSG-Cut* operation resulted in a single solid.

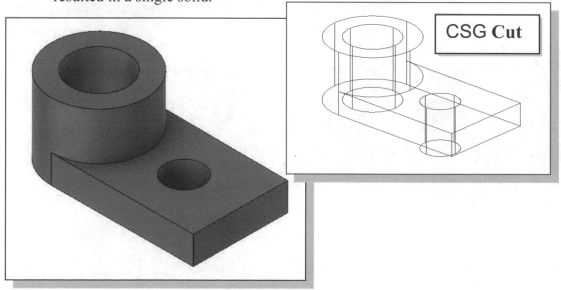

Creating a Rectangular Cut Feature

- Next create a rectangular cut as the last solid feature of the *Locator*.

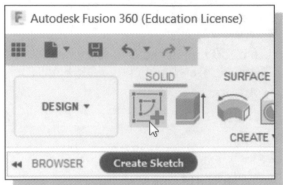

1. In the *Create toolbar* select the **Create Sketch** command by left-clicking once on the icon.

2. Pick the right face of the base feature as shown.

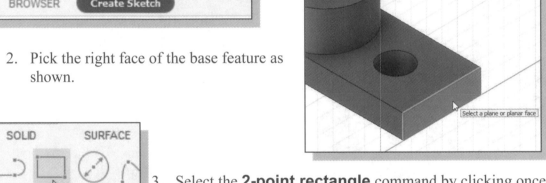

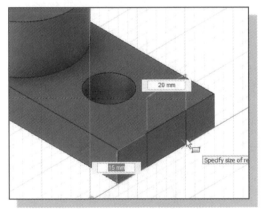

3. Select the **2-point rectangle** command by clicking once with the left-mouse-button on the icon in the *Sketch tab* of the Sketch toolbar.

4. Create a rectangle that is aligned to the top and bottom edges of the base feature as shown. (Hint: first set the display to isometric orientation.)

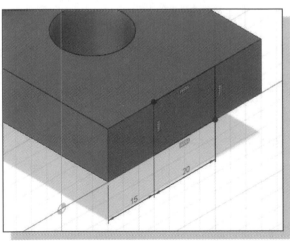

5. On your own, create and modify the two dimensions (15 & 20 mm) as shown.

6. Select **Finish Sketch** in the *Ribbon* toolbar to end the **Sketch** option.

7. In the *Create toolbar*, select the **Extrude** command by left-clicking on the icon.

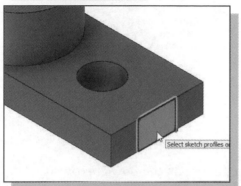

8. In the *Extrude* pop-up window, the **Profile** button is pressed down; Autodesk Fusion 360 expects us to identify the profile to be extruded. Select the rectangle region as the profile to be extruded.

9. In the *Extrude* pop-up window, set the operation option to **Cut**. Select **To object** as the *Extents* option and choose the small circular cut feature as the termination object.

10. Click on the **OK** button to create the *Cut* feature and complete the design.

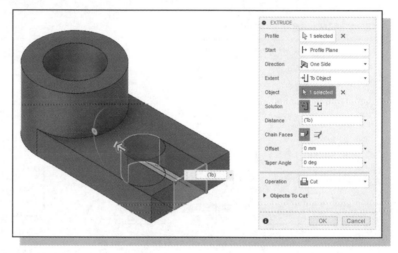

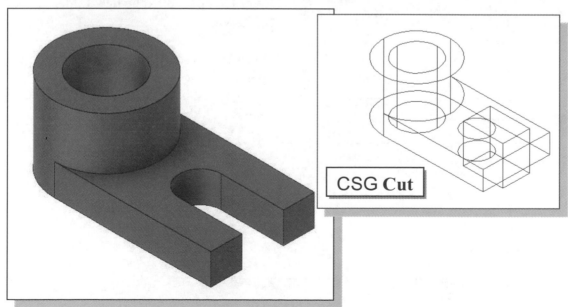

CSG Cut

Save the Model

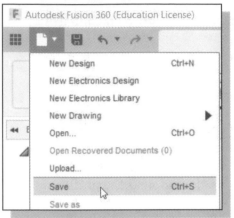

1. Select **Save** in the *Quick Access* toolbar, or you can also use the "**Ctrl-S**" combination (hold down the "**Ctrl**" key and hit the "**S**" key once) to save the part.

2. Expand the *option* list by clicking on the *down arrow* as shown.

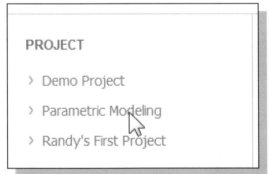

3. Select the **Parametric Modeling** Project by clicking once with the left mouse button.

4. Click **New Folder** to create a new subfolder.

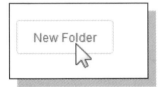

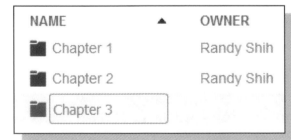

5. Enter **Chapter3** as the new folder name as shown.

6. **Double-click** on the Chapter 3 folder to open it.

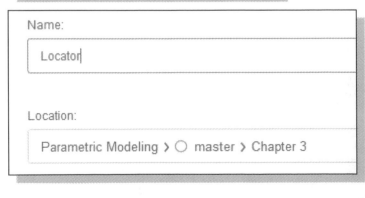

7. In the *file name* editor box, enter **Locator** as the file name.

8. Click on the **Save** button to save the file.

Review Questions:

1. List and describe three basic *Boolean operations* commonly used in computer geometric modeling software.

2. What is a *primitive solid*?

3. What does *CSG* stand for?

4. Which *Boolean operation* keeps only the volume common to the two solid objects?

5. What is the main difference between a *CUT feature* and a *HOLE feature* in Autodesk Fusion 360?

6. Create the following 2D Sketch and measure the associated area and perimeter.

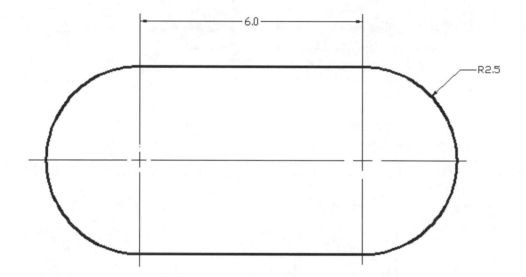

7. Using the CSG concepts, create *Binary Tree* sketches showing the steps you plan to use to create the two models shown on page 3-29:

Exercises: Create and save the exercises in the Chapter3 folder.

1. **Latch Clip** (Dimensions are in inches. Thickness: **0.25** inches.)

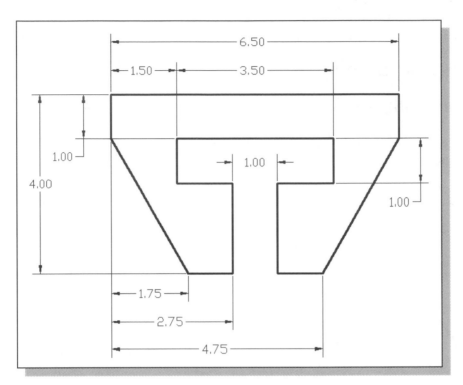

2. **Guide Plate** (Dimensions are in inches. Thickness: **0.25** inches. Boss height **0.125** inches.)

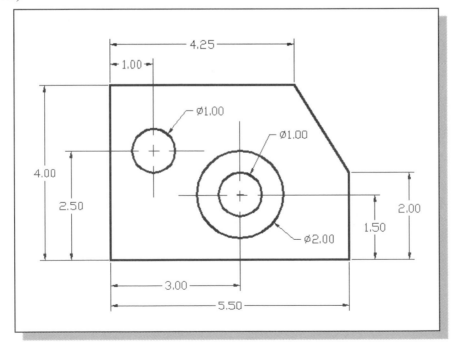

3. **Angle Slider** (Dimensions are in Millimeters.)

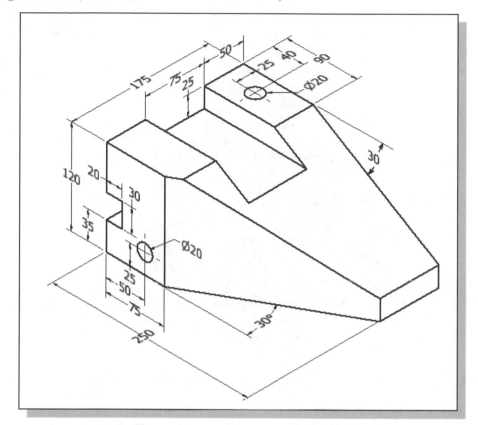

4. **Coupling Base** (Dimensions are in inches.)

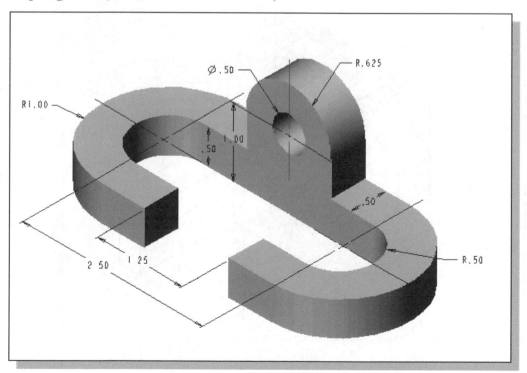

5. **Indexing Guide** (Dimensions are in inches.)

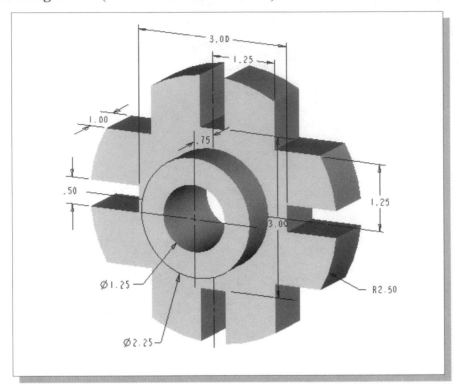

6. **L-Bracket** (Dimensions are in inches.)

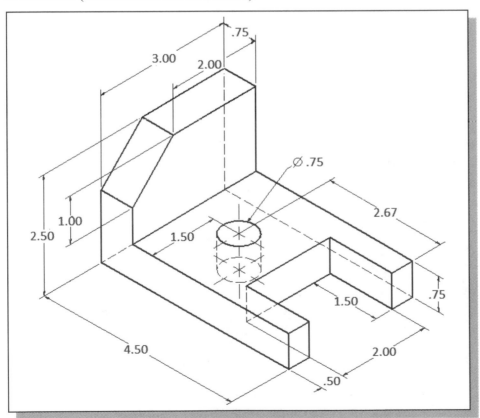

Chapter 4
Model History Tree

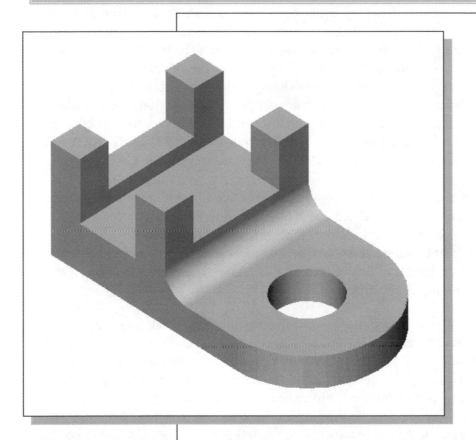

Learning Objectives

- ♦ **Understand Feature Interactions**
- ♦ **Use the Part Browser**
- ♦ **Modify and Update Feature Dimensions**
- ♦ **Perform History-Based Part Modifications**
- ♦ **Change the Names of Created Features**
- ♦ **Implement Basic Design Changes**

Introduction

In Autodesk Fusion 360, the **design intents** are embedded into features in the **history tree**. The structure of the model history tree resembles that of a **CSG binary tree**. A CSG binary tree contains only *Boolean relations*, while the **Autodesk Fusion 360 history tree** contains all features, including *Boolean relations*. A history tree is a sequential record of the features used to create the part. This history tree contains the construction steps, plus the rules defining the design intent of each construction operation. In a history tree, each time a new modeling event is created, previously defined features can be used to define information such as size, location, and orientation. It is therefore important to think about your modeling strategy before you start creating anything. This approach in modeling is a major difference of **FEATURE-BASED CAD SOFTWARE**, such as Autodesk Fusion 360, from previous generation CAD systems.

Feature-based parametric modeling is a cumulative process. Every time a new feature is added, a new result is created and the feature is also added to the history tree. The database also includes parameters of features that were used to define them. All of this happens automatically as features are created and manipulated. At this point, it is important to understand that all of this information is retained, and modifications are done based on the same input information.

In Autodesk Fusion 360, the history tree gives information about modeling order and other information about the feature. Part modifications can be accomplished by accessing the features in the history tree. It is therefore important to understand and utilize the feature history tree to modify designs. Autodesk Fusion 360 remembers the history of a part, including all the rules that were used to create it, so that changes can be made to any operation that was performed to create the part. In Autodesk Fusion 360, the history tree can be accessed through the ***Browser*** window on the left and the ***Timeline*** at the bottom of the main window.

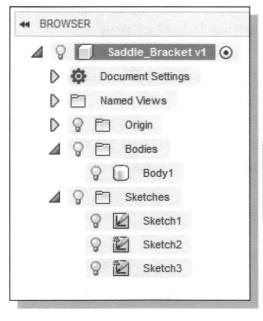

Browser and **Timeline**:
Sequential record of the construction steps

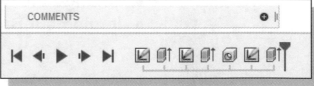

The Saddle Bracket Design

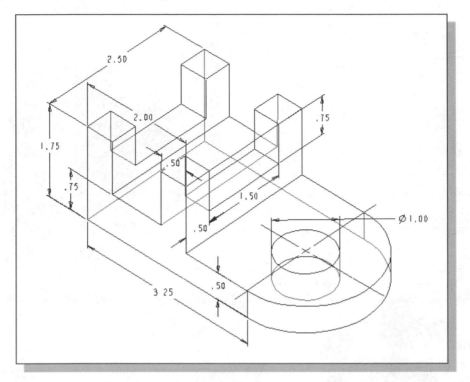

❖ Based on your knowledge of Autodesk Fusion 360 so far, how many features would you use to create the design? Which feature would you choose as the **BASE FEATURE**, the first solid feature, of the model? What is your choice in arranging the order of the features? Would you organize the features differently if additional fillets were to be added in the design? Take a few minutes to consider these questions and do preliminary planning by sketching on a piece of paper. You are also encouraged to create the model on your own prior to following through the tutorial.

Starting Autodesk Fusion 360

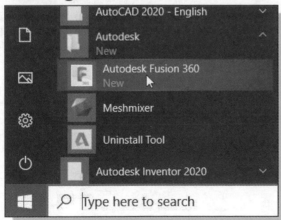

1. Select the **Autodesk Fusion 360** option on the *Start* menu or select the **Autodesk Fusion 360** icon on the desktop to start Autodesk Fusion 360.

2. In the *Sign In* dialog box, log in with your email or username.

Modeling Strategy

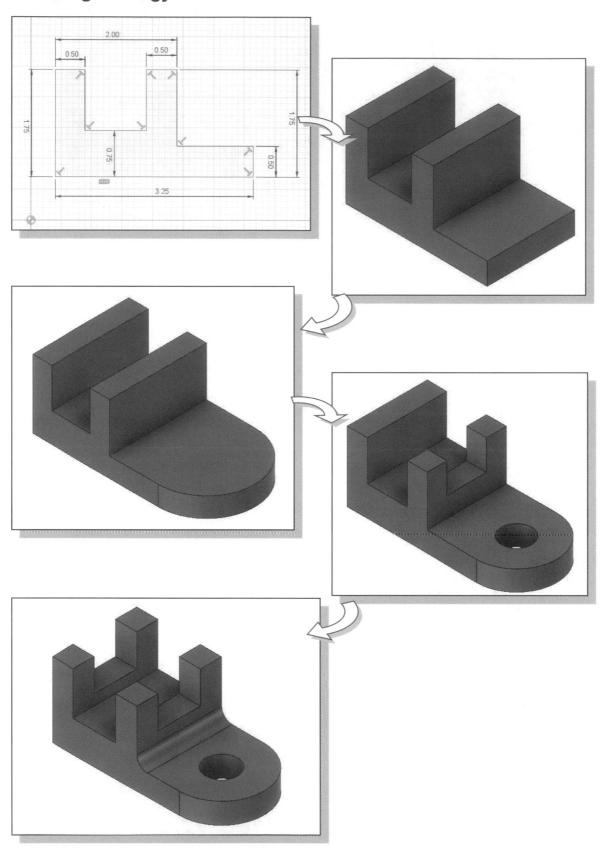

The Autodesk Fusion 360 Browser and Timeline Control

In the Autodesk Fusion 360 screen layout, the **browser** is located to the left of the graphics window. Autodesk Fusion 360 can be used for part modeling, assembly modeling, part drawings, and assembly presentation. The *browser* window provides a visual structure of the features, constraints, and attributes that are used to create the part, assembly, or scene. The *browser* also provides right-click menu access for tasks associated specifically with the part or feature, and it is the primary focus for executing many of the Autodesk Fusion 360 commands.

The Timeline control panel is located at the bottom of the main window. Each time a new Fusion 360 feature is created, the feature is added to the Fusion 360 Timeline. Features in the timeline are identified by the type of operations. The timeline can be re-played or rolled back to any specific point of the feature creations. Both the *browser* and *timeline* can also be used to modify parts and assemblies by moving, deleting, or renaming items within the hierarchy. Any changes made in the *browser* directly affect the part or assembly and the results of the modifications are displayed on the screen instantly.

Create the Base Feature

1. Expand the **Document Setting** item in the Browser area to show the current units setup.

2. On your own, set the units to **inches** as shown.

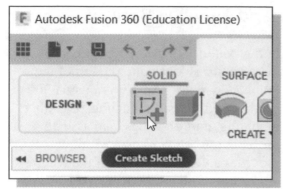

3. Activate the **Create Sketch** icon with a single click of the left-mouse-button.

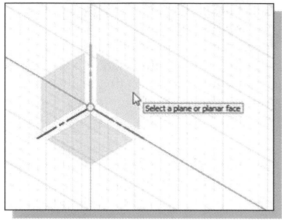

4. Move the cursor over the right-vertical plane (the *XY Plane)* in the graphics area. Click once with the **left-mouse-button** to select the *Plane* as the sketch plane for the new sketch.

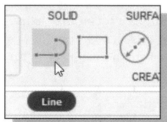

5. Select the **Line** icon by clicking once with the left-mouse-button; this will activate the Line command.

6. On your own, create and adjust the geometry by adding and modifying dimensions as shown below.

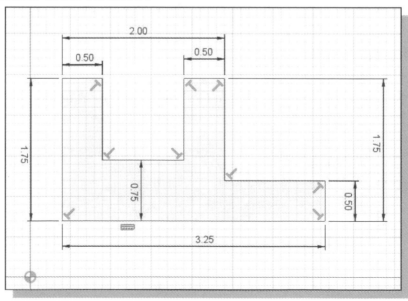

7. In the *Ribbon* toolbar, select **Finish Sketch** to exit the Sketch mode.

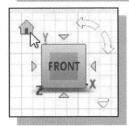

8. On your own, use the dynamic viewing functions to view the sketch. Click the home view icon to change the display to the *isometric* view.

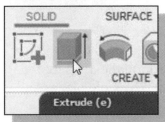

9. In the *Create toolbar*, select the **Extrude** command by left-clicking on the icon.

10. Notice the 2D sketch is pre-selected as the sketch profile to be extruded.

11. In the *Distance* option box, enter **2.5** as the total extrusion distance.

12. In the *Extrude* pop-up window, select **Symmetric** using the **Whole Length** option. The **Symmetric** option allows us to extrude in both directions of the sketched profile.

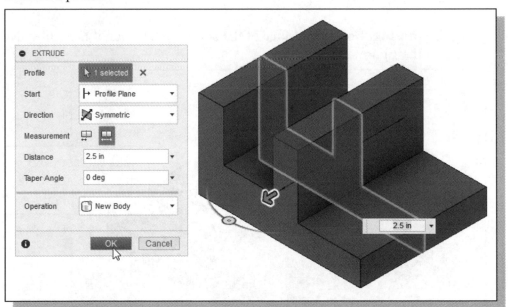

13. Click on the **OK** button to accept the settings and create the base feature.

➢ On your own, use the *Dynamic Viewing* functions to view the 3D model. Also, note the extrusion feature is added to the *Timeline* and the *browser* area.

Create the Second Solid Feature

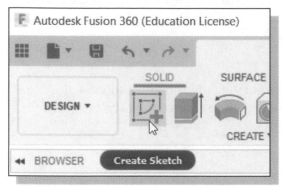

1. In the *Sketch toolbar* select the **Create Sketch** command by left-clicking once on the icon.

2. In the *graphics* area, the message "*Select a plane or planar surface*" is displayed. Autodesk Fusion 360 expects us to identify a planar surface where the 2D sketch of the next feature is to be created. Move the graphics cursor on the 3D part and notice that Autodesk Fusion 360 will automatically highlight feasible planes and surfaces as the cursor is on top of the different surfaces.

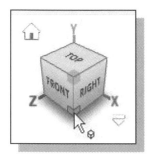

3. Click the bottom corner of the View Cube to rotate the display to view the bottom surface of the model.

4. Select the **bottom horizontal face** of the solid model when it is highlighted as shown in the figure.

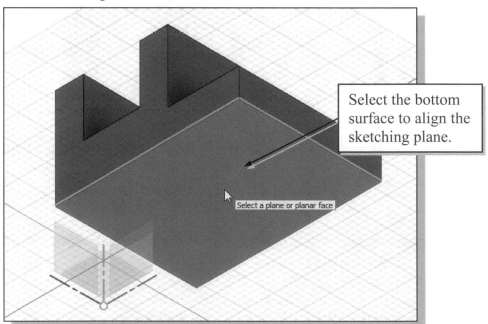

Select the bottom surface to align the sketching plane.

Select a plane or planar face

Create a 2D Sketch

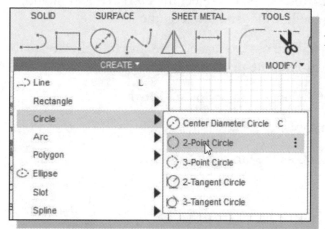

1. Select the **2-Point Circle** command by clicking once with the left-mouse-button on the icon in the *Sketch* panel.

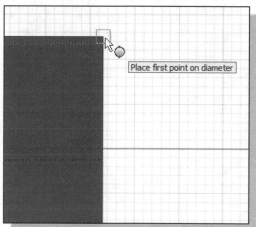

➢ We will align the circle to the two corners of the base feature.

2. Pick the **top corner** of the surface as shown in the figure.

3. Pick the **bottom corner** of the surface as shown in the figure.

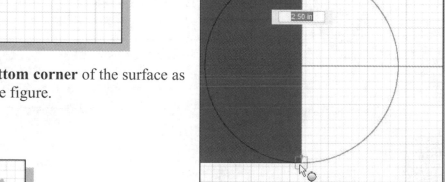

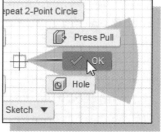

4. Inside the graphics window, click once with the right-mouse-button to display the option menu. Select **OK** in the pop-up menu to end the Circle command.

5. Inside the graphics window, click once with the right-mouse-button to display the option menu. Select **Finish Sketch** in the pop-up menu to end the Sketch option.

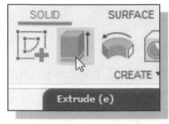

6. In the *Create* toolbar panel, select the **Extrude** command by clicking the left-mouse-button on the icon.

7. Click inside the **circle** we just created and left-click once to select the region as the **profile** to be extruded.

8. In the *Extrude* pop-up control, set the operation option to **Join.**

9. Also, set the *Extents* option to **To Object** as shown below.

10. Select the top face of the base feature as the termination surface for the extrusion.

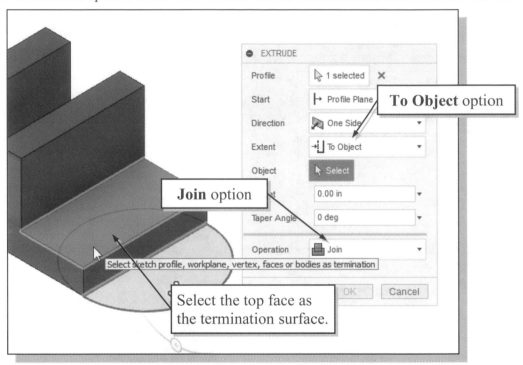

11. Set the *Chain Faces* option to **Extend Faces** as shown.

12. Click on the **OK** button to proceed with the *Solid feature*.

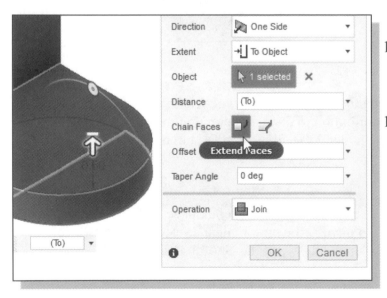

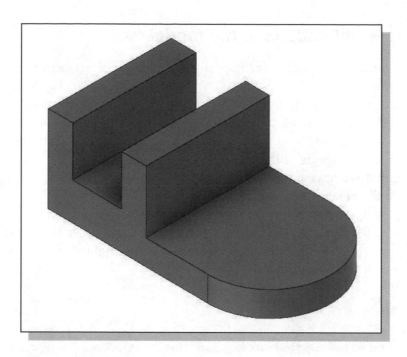

Renaming the Part Features

Currently, our model contains two extruded features. The feature is highlighted in the display area when we select the feature in the *browser* window. Each time a new feature is created, the feature is also displayed in the *Model Tree* window. By default, Autodcsk Fusion 360 will use generic names for part features. However, when we begin to deal with parts with a large number of features, it will be much easier to identify the features using more meaningful names.

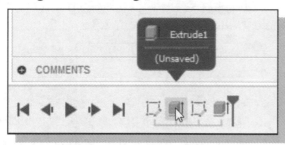

1. Select the first extruded feature in the *model browser* area by left-clicking once on the name of the feature, **Extrude1**. Notice the selected feature is highlighted in the graphics window.

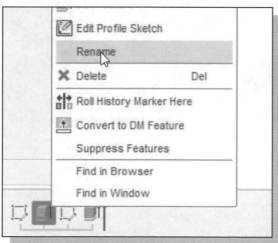

2. Right-click once on the feature to bring up the option list and choose **Rename** as shown.

3. Enter **Base** as the new name for the first extruded feature.

4. On your own, rename the second extruded feature to **Circular_End**.

History Based Modifications to the model

One of the main advantages of parametric modeling is the ease of performing part modifications at any time in the design process. Part modifications can be done through accessing the features in the history tree. Autodesk Fusion 360 remembers the history of a part, including all the rules that were used to create it, so that changes can be made to any operation that was performed to create the part. We can think of it as going back in time and modifying some aspects of the modeling steps used to create the part. We can modify any feature that we have created. For our *Saddle Bracket* design, we will first reduce the extrusion distance of the base feature.

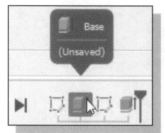

1. Select the first extruded feature, **Base**, in the *Timeline* area. Notice the selected feature is highlighted in the graphics window.

2. Inside the *Timeline* area, **right-click** on the first extruded feature to bring up the option menu and select the **Edit Feature** option in the pop-up menu.

3. In the *Edit Feature* dialog box, change the extruded distance from **2.5** to **2.0.**

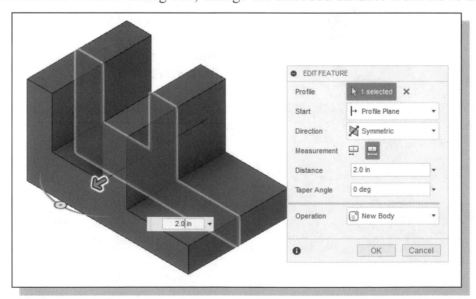

4. Click **OK** to accept the settings.

Adding a Placed Feature

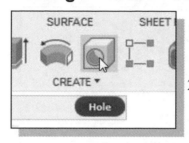

1. In the *Create toolbar*, select the **Hole** command by left-clicking on the icon.

2. Pick the **top plane** of the round end as the placement plane as shown.

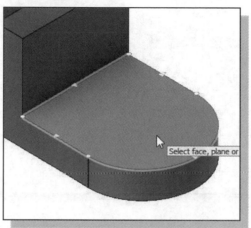

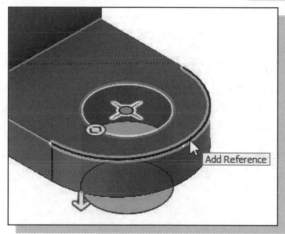

3. Click on the arc (*Add Reference*); this will add a **Concentric** constraint to place the center point of the hole feature.

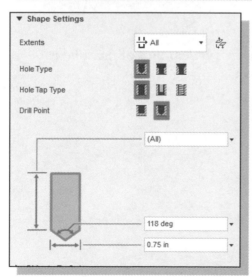

4. Set the termination *Extents* option to **Through All** as shown.

5. Set the hole diameter to **0.75 in** as shown.

6. Click **OK** to accept the settings and create the *Hole* feature.

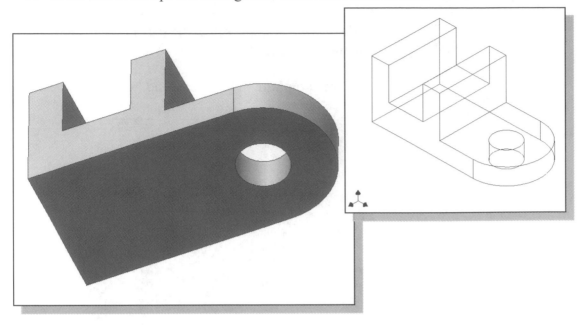

Create a Rectangular Cut Feature

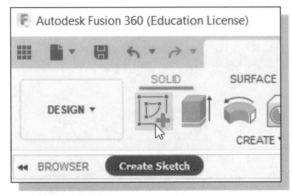

1. In the *Create toolbar* select the **Create Sketch** command by left-clicking once on the icon.

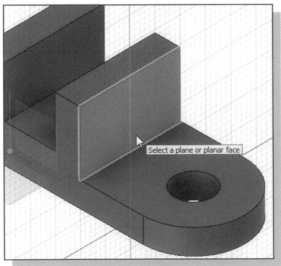

2. Pick the **vertical face** of the solid as shown.

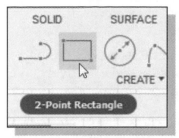

3. Activate the **2-point rectangle** command by clicking once with the left-mouse-button on the icon in the *Sketch* toolbar.

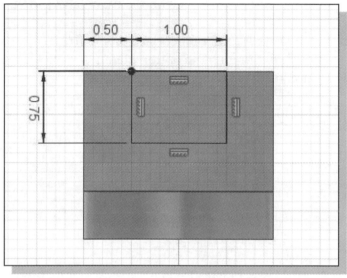

4. On your own, create a rectangle (size **1.0 × 0.75**) as shown.

➤ On your own, create a rectangular cut feature (depth: 0.5 in) as shown and rename the feature to **Rect_Cut**.

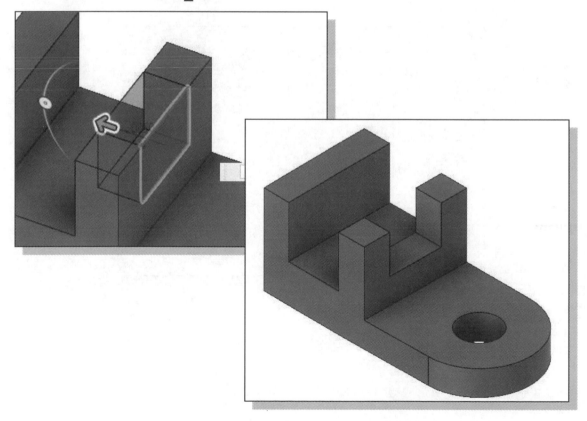

A Design Change – Edit the 2D sketches

Engineering designs usually go through many revisions and changes. Autodesk Fusion 360 provides an assortment of tools to handle design changes quickly and effectively. We will demonstrate some of the tools available by changing the **Base** feature of the design.

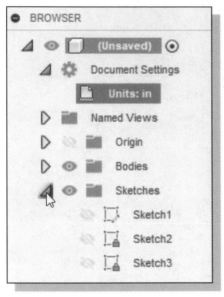

1. In the *browser* window, expand the **Sketches** item by left-clicking once on the triangle as shown.

2. In the *browser*, right-click once on the **first sketch** to bring up the option menu; then pick **Edit Sketch** in the pop-up menu.

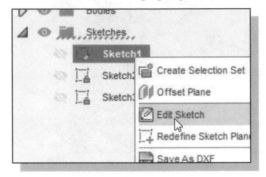

❖ Autodesk Fusion 360 will now display the original 2D sketch of the first feature in the graphics window. We have literally gone back to the point where we first created the 2D sketch. Notice the sketch being modified is also highlighted in the desktop *browser*.

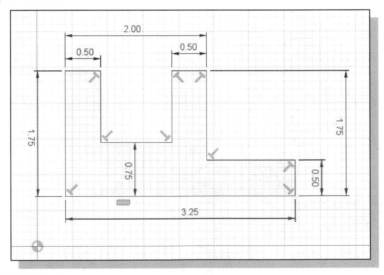

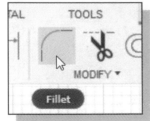

3. Select the **Fillet** command in the *2D Sketch* toolbar.

4. Select the inside corner on the left as shown to add a rounded fillet.

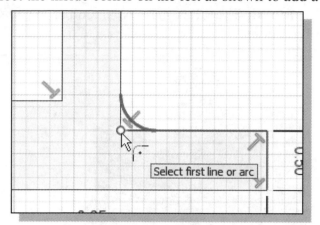

5. In the graphics window, enter **0.25** as the new radius of the fillet.

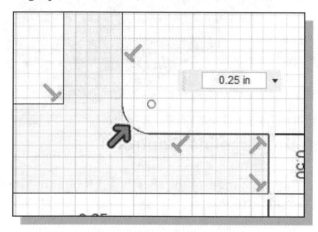

- Note that the fillet is created automatically with the dimension attached. The attached dimension can also be modified through the history tree.

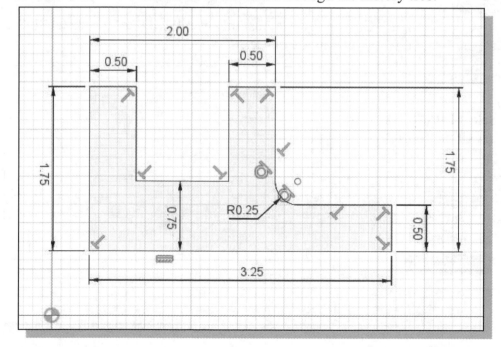

6. Select **Finish Sketch** in the *Ribbon* toolbar to end the Sketch option.

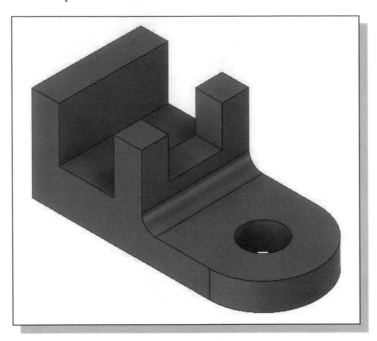

Edit the Sketch through the Timeline Control

*   Autodesk Fusion 360 also allows us to modify the sketches through the timeline control near the bottom of the main window.

1. Double click on the **first sketch** (**Sketch1**) in the Timeline control.

2. On your own, adjust the overall width of the sketch to **3.0** as shown in the figure.

3. Select **Finish Sketch** to exit the Sketch option.

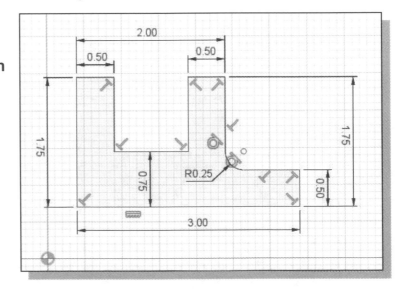

Direct Part Modifications

- Autodesk Fusion 360 also provides the *Direct part modification* option, which enables us to make modifications directly on the solid model. As an example, we will adjust the depth of the rectangular cutout.

    1. In the *Graphics* window, right-click once on one of the vertical surfaces of the **Rect_Cut** feature to display the pop-up menu. Note that we can also edit the associated 2D sketch as well.

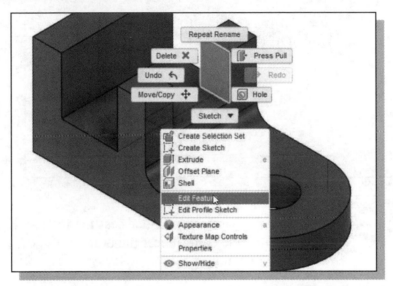

    2. Select **Edit Feature** in the pop-up menu. Notice the *Edit Feature* dialog box appears on the screen.

    3. In the *Edit Feature* dialog box, set the termination *Extent* to the **Through All** option.

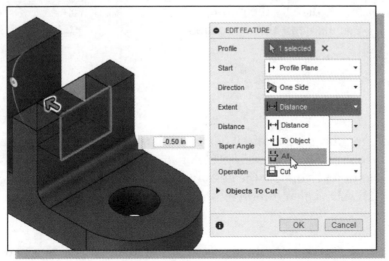

    4. Click on the **OK** button to accept the settings.

- As can been seen, the history-based modification approach is very straightforward, and it only took a few seconds to adjust the cut feature to the **Through All** option.

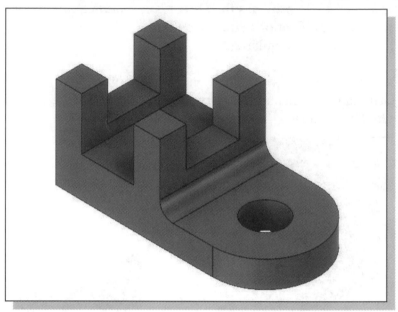

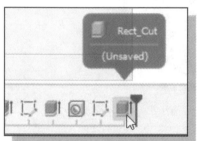

5. Note that we can also access the feature through the Timeline control near the bottom of the Fusion 360 main window.

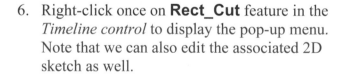

6. Right-click once on **Rect_Cut** feature in the *Timeline control* to display the pop-up menu. Note that we can also edit the associated 2D sketch as well.

❖ In a typical design process, the initial design will undergo many analyses, testing, and reviews. The *history-based part modification* approach is an extremely powerful tool that enables us to quickly update the design. At the same time, it is quite clear that PLANNING AHEAD is also important in doing feature-based modeling.

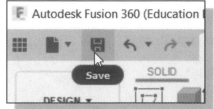

7. In the *Standard* toolbar, click **Save** and save the model as *Saddle_Bracket* in the Chapter4 folder.

Assigning and Calculating the Associated Physical Properties

Autodesk Fusion 360 models can have physical properties. The associated *properties* can be used in bills of materials and drawing parts lists. We can also set and calculate physical properties for a part or assembly using the material library. This allows us to examine the physical properties of the model, such as weight or center of gravity.

1. In the *Modify panel*, select **Physical Material** as shown.

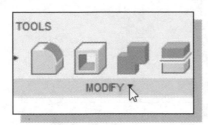
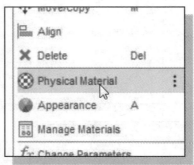

2. By default, the Steel material is available in the ***Physical Material*** dialog box.

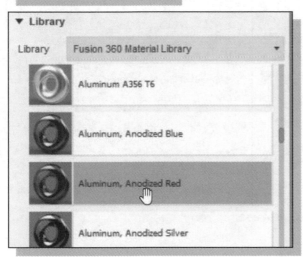

3. On your own, look at the different type of materials available in the ***Physical Material*** dialog box.

4. Scroll down in the material lists and select **Aluminum, Anodized Red** under the **Metal** folder by *Drag and Drop* to the *In This Design* window as shown.

5. To assign the material to the part, **drag and drop** the material to the 3D model inside the graphics window.

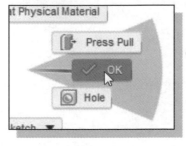

6. Close the ***Physical Material*** dialog box by right-clicking once and click **OK** as shown.

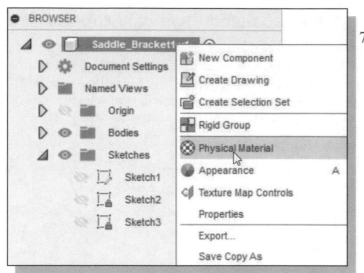

7. In the browser window, select the part name with the right-mouse-button-click to display the option menu. Select **Properties** as shown.

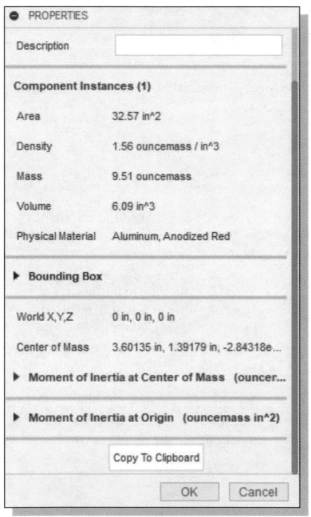

❖ Note the *Properties* dialog box shows detailed information regarding the created model, such as the *Mass*, *Area*, *Volume* and the *Center of Gravity* information of the model based on the density of the selected material. Note that Moment of Inertia information is also available.

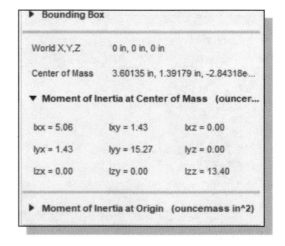

8. On your own, select **Cast Iron** as the *Material* type and compare the differences in using the different materials.

Review Questions:

1. What are stored in the Autodesk Fusion 360 *History Tree*?

2. When extruding, what is the difference between *Distance* and *Through All*?

3. Describe the *history-based part modification* approach.

4. What determines how a model reacts when other features in the model change?

5. Describe the steps to rename existing features.

6. Describe two methods available in Autodesk Fusion 360 to *modify the dimension values* of parametric sketches.

7. Create *History Tree sketches* showing the steps you plan to use to create the two models shown on the next page:

Ex.1)

Ex.2)

Exercises: Create and save the exercises in the Chapter4 folder.

1. **C-Clip** (Dimensions are in inches. Plate thickness: **0.25 inches**.)

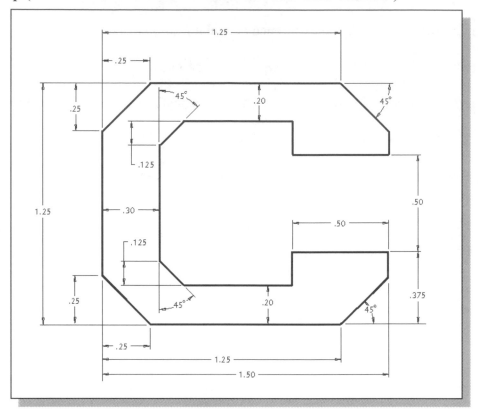

2. **Tube Mount** (Dimensions are in inches.)

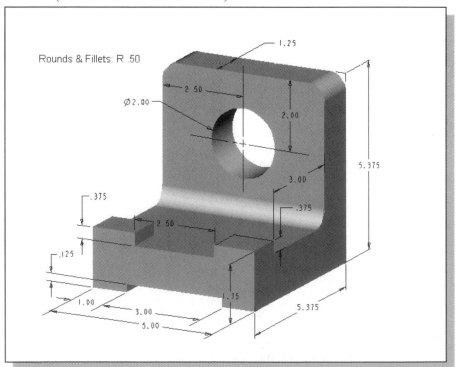

3. **Hanger Jaw** (Dimensions are in inches. Volume =?)

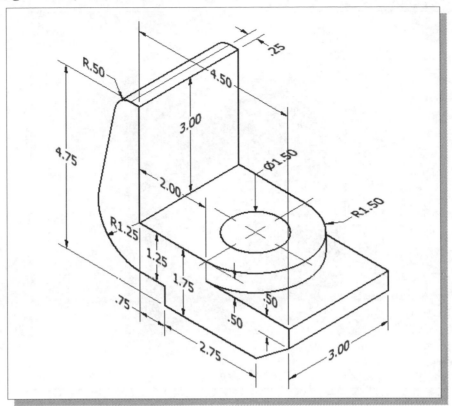

4. **Transfer Fork** (Dimensions are in inches. Material: **Cast Iron.** Volume =?)

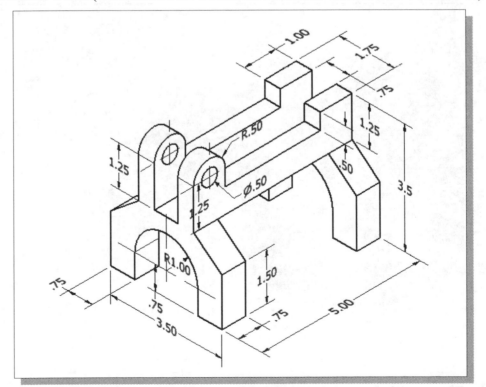

5.   **Guide Slider** (Material: **Cast Iron**. Weight and Volume =?)

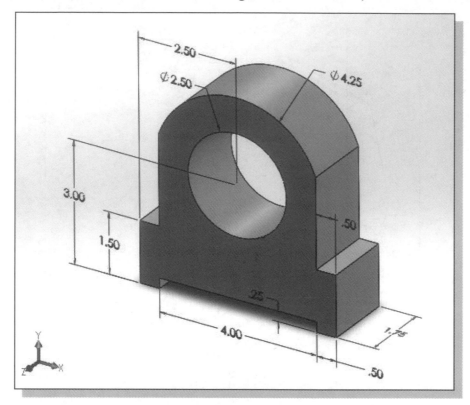

6.   **Shaft Guide** (Material: **Aluminum-6061**. Mass and Volume =?)

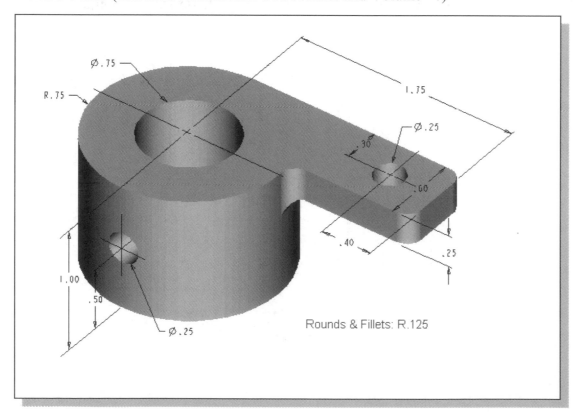

Chapter 5
Parametric Constraints Fundamentals

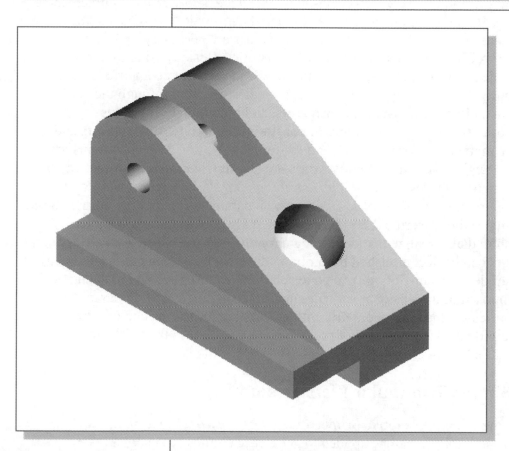

Learning Objectives

- ♦ **Create Parametric Relations**
- ♦ **Use Dimensional Variables**
- ♦ **Display, Add, and Delete Geometric Constraints**
- ♦ **Understand and Apply Different Geometric Constraints**
- ♦ **Display and Modify Parametric Relations**
- ♦ **Create Fully Constrained Sketches**

CONSTRAINTS and RELATIONS

A primary and essential difference between parametric modeling and previous generation computer modeling is that parametric modeling captures the *design intent*. In the previous lessons, we have seen that the design philosophy of *"shape before size"* is implemented through the use of Autodesk Fusion 360's Profile and Dimension commands. In performing geometric constructions, dimensional values are necessary to describe the **SIZE** and **LOCATION** of constructed geometric entities. Besides using dimensions to define the geometry, we can also apply geometric rules to control geometric entities. More importantly, Autodesk Fusion 360 can capture design intent through the use of **geometric constraints**, **dimensional constraints**, and **parametric relations**. In Autodesk Fusion 360, there are two types of constraints: **geometric constraints** and **dimensional constraints**. For part modeling in Autodesk Fusion 360, constraints are applied to *2D sketches*. **Geometric constraints** are **geometric restrictions** that can be applied to geometric entities; for example, *horizontal*, *parallel*, *perpendicular*, and *tangent* are commonly used *geometric constraints* in parametric modeling. **Dimensional constraints** are used to describe the SIZE and LOCATION of individual geometric shapes. One should also realize that depending upon the way the constraints are applied, the same results can be accomplished by applying different constraints to the geometric entities. In Autodesk Fusion 360, **parametric relations** are user-defined mathematical equations composed of dimensional variables and/or *design variables*. In parametric modeling, features are made of geometric entities with both relations and constraints describing individual design intent. In this lesson, we will discuss the fundamentals of parametric relations and geometric constraints.

Create a Simple Triangular Plate Design

In parametric modeling, **geometric properties** such as *horizontal*, *parallel*, *perpendicular*, and *tangent* can be applied to geometric entities automatically or manually. By carefully applying proper **geometric constraints**, very intelligent models can be created. This concept is illustrated by the following example.

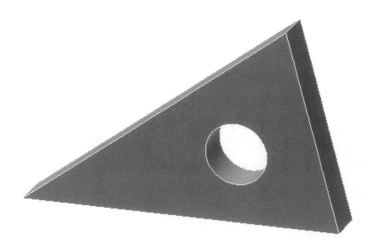

Fully Constrained Geometry

In Autodesk Fusion 360, as we create 2D sketches, geometric constraints such as *horizontal* and *parallel* are automatically added to the sketched geometry. In most cases, additional constraints and dimensions are needed to fully describe the sketched geometry beyond the geometric constraints added by the system. Although we can use Autodesk Fusion 360 to build partially constrained or totally unconstrained solid models, the models may behave unpredictably as changes are made. In most cases, it is important to consider the design intent and to add proper constraints to geometric entities. In the following sections, a simple triangle is used to illustrate the different tools that are available in Autodesk Fusion 360 to create/modify geometric and dimensional constraints.

Starting Autodesk Fusion 360

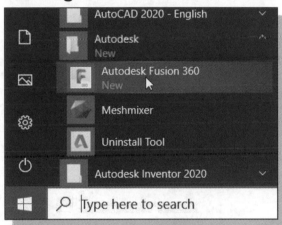

1. Select the **Autodesk Fusion 360** option on the *Start* menu or select the **Autodesk Fusion 360** icon on the desktop to start Autodesk Fusion 360.

2. In the *Sign In* dialog box, log in with your email or username.

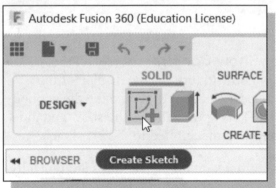
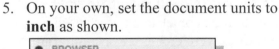

3. In the *Sketch panel,* click once with the left-mouse-button to select the **Create Sketch** command.

4. Click once with the **left-mouse-button** to select the right-vertical plane (the *XY Plane)* as the sketch plane for the new sketch.

5. On your own, set the document units to **inch** as shown.

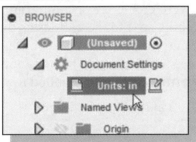

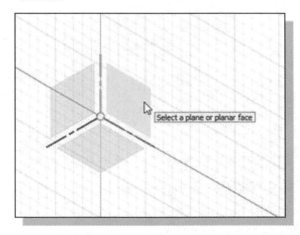

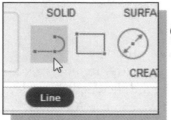

6. Click the **Line** icon in the *Sketch* tab to activate the command. A Help-tip box appears next to the cursor, and a brief description of the command is displayed at the bottom of the drawing screen: "Creates Straight line and arcs."

7. Create a triangle of arbitrary size positioned near the center of the screen as shown below. (Note that the base of the triangle is horizontal.)

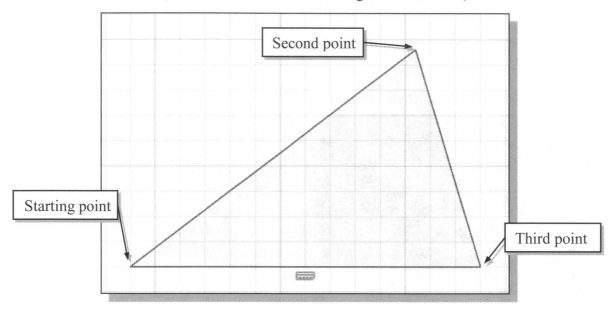

Display and Hide Existing Constraints

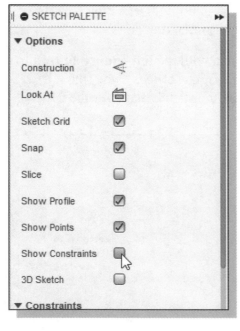

1. Click the **Show Constraints** icon in the *Sketch Palette* window to turn off the display of the constraints. Click again to re-display the applied constraint. This option allows the show and hide constraints that are applied to the 2D sketch.

➢ In *parametric modeling*, constraints are typically applied as geometric entities are created. Autodesk Fusion 360 will attempt to add proper constraints to the geometric entities based on the way the entities were created. Constraints are displayed as symbols next to the entities as they are created. The current profile consists of three line entities, three straight lines. The horizontal line has a **Horizontal** *constraint* applied to it.

Autodesk Fusion 360 Geometric Constraints

In Autodesk Fusion 360, thirteen types of constraints are available for 2D sketches. Constraints relate sketch entities to one another geometrically. This helps the sketch entities maintain certain behaviors when the sketch updates. All of the constraints can be accessed from the Sketch Palette.

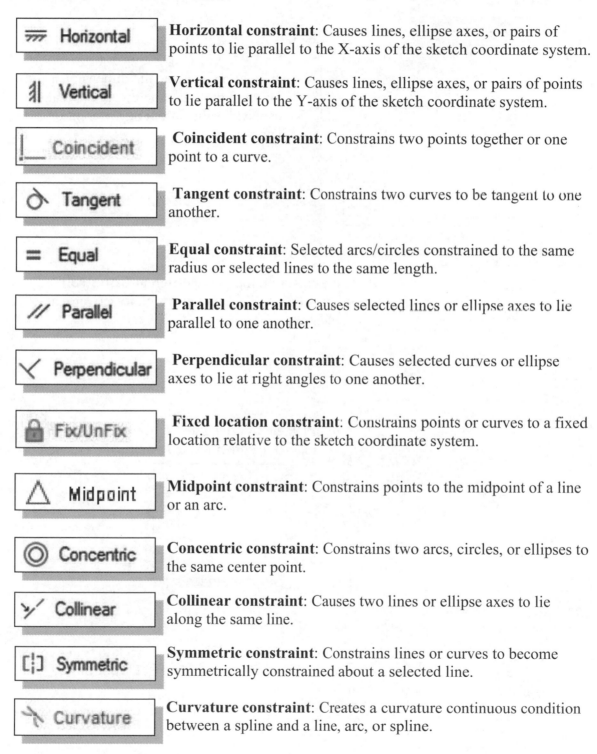

Horizontal constraint: Causes lines, ellipse axes, or pairs of points to lie parallel to the X-axis of the sketch coordinate system.

Vertical constraint: Causes lines, ellipse axes, or pairs of points to lie parallel to the Y-axis of the sketch coordinate system.

Coincident constraint: Constrains two points together or one point to a curve.

Tangent constraint: Constrains two curves to be tangent to one another.

Equal constraint: Selected arcs/circles constrained to the same radius or selected lines to the same length.

Parallel constraint: Causes selected lines or ellipse axes to lie parallel to one another.

Perpendicular constraint: Causes selected curves or ellipse axes to lie at right angles to one another.

Fixed location constraint: Constrains points or curves to a fixed location relative to the sketch coordinate system.

Midpoint constraint: Constrains points to the midpoint of a line or an arc.

Concentric constraint: Constrains two arcs, circles, or ellipses to the same center point.

Collinear constraint: Causes two lines or ellipse axes to lie along the same line.

Symmetric constraint: Constrains lines or curves to become symmetrically constrained about a selected line.

Curvature constraint: Creates a curvature continuous condition between a spline and a line, arc, or spline.

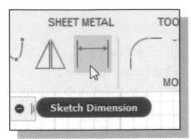

1. In the ribbon toolbar, select the Sketch Dimension command as shown. (You can also hit the [**D**] key.) The Sketch Dimension command allows us to quickly create and modify dimensions.

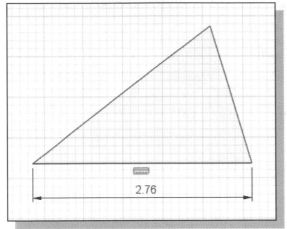

2. On your own, create the dimension as shown in the figure below. (Note that the displayed value might be different on your screen.)

3. Move the cursor on top of the different **Constraint** icons. Click on the **Fix** constraint icon to activate the command.

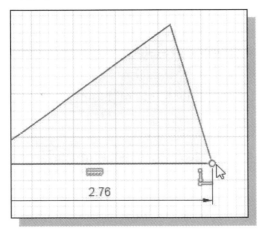

4. Pick the **lower right corner** of the triangle to make the corner a fixed point.

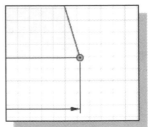

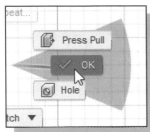

5. Inside the graphics window, right-mouse-click to bring up the option menu and select **OK** to end the Fix Constraints command.

➢ Geometric constraints can be used to control the direction in which changes can occur. For example, in the current design we are adding a horizontal dimension to control the length of the horizontal line. If the length of the line is modified to a greater value, Autodesk Fusion 360 will lengthen the line toward the left side. This is due to the fact that the Fix constraint will restrict any horizontal movement of the horizontal line toward the right side.

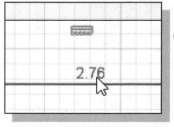

6. Double click on the dimension text to open the *Edit Dimension* window.

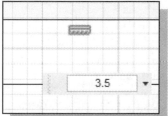

7. Enter a value that is greater than the displayed value to observe the effects of the modification. (For example, if the dimension value is 2.76, then enter **3.5** in the text edit box.)

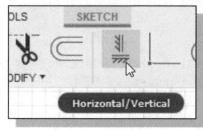

8. On your own, use the **Undo** command to reset the dimension value to the previous value.

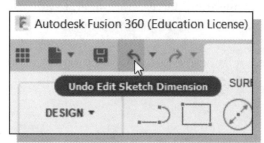

9. Select the **Horizontal/Vertical** constraint icon in the *Sketch Palette* window.

10. Pick the inclined line on the right to make the line vertical as shown in the figure below.

11. Hit the [**Esc**] key once to end the Vertical Constraint command.

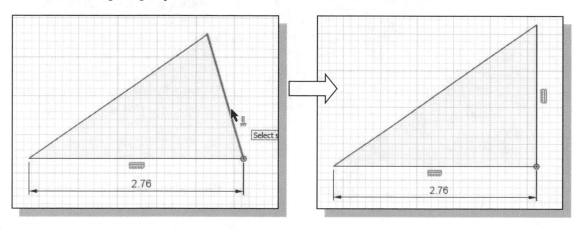

➤ One should think of the constraints and dimensions as defining elements of the geometric entities. How many more constraints or dimensions will be necessary to fully constrain the sketched geometry? Which constraints or dimensions would you use to fully describe the sketched geometry?

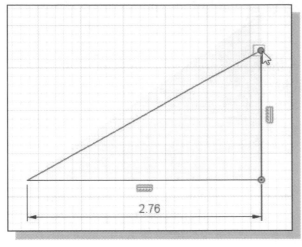

12. Drag the top corner of the triangle and note that the corner can be moved to a new location. Release the mouse button at a new location and notice the corner is adjusted only in an upward or downward direction. Note that the two adjacent lines are automatically adjusted to the new location.

13. On your own, experiment with dragging the other corners to new locations.

• The **Vertical** constraint, along with the **Fix** constraint at the lower right corner, does not fully describe the location of the top corner of the triangle. We will need to add additional information, such as the length of the vertical line or an angle dimension.

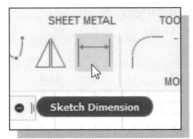

14. In the Create Panel, select the **Sketch Dimension** command as shown. (You can also hit the [D] key.)

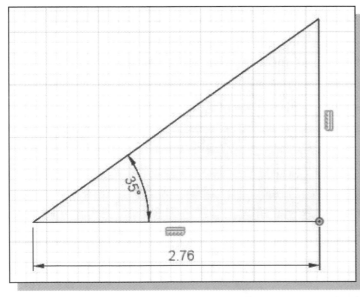

15. On your own, add an angle dimension to the left corner of the triangle.

16. Press the [**Esc**] key once to exit the **Sketch Dimension** command.

17. On your own, modify the angle to **35°** and then experiment with dragging the top corner of the triangle.

• Note the sketched geometry is fully constrained with the added dimension.

Over-Constraining and Driven Dimensions

We can use Autodesk Fusion 360 to build partially constrained or totally unconstrained solid models. In most cases, these types of models may behave unpredictably as changes are made. However, Autodesk Fusion 360 will not let us over-constrain a sketch; additional dimensions can still be added to the sketch, but they are used as references only. These additional dimensions are called **_driven dimensions_**. _Driven dimensions_ do not constrain the sketch; they only reflect the values of the dimensioned geometry. They are enclosed in parentheses to distinguish them from normal (parametric) dimensions. A _driven dimension_ can be converted to a normal dimension only if another dimension or geometric constraint is removed.

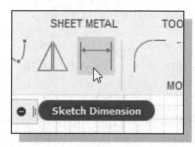

1. Select the **Sketch Dimension** command in the _Sketch Panel_.

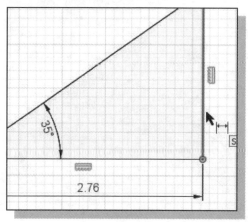

2. Select the **vertical line**.

3. Pick a location that is to the right side of the triangle to place the dimension text.

4. A warning dialog box appears on the screen stating that the dimension we are trying to create will over-constrain the sketch. Click on the **OK** button to proceed with the creation of a driven dimension. Note the brackets around the _Driven Dimension_.

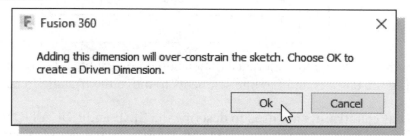

5. On your own, modify the angle dimension to **30°** and observe the changes of the 2D sketch and the driven dimension. Change the dimension back to **35°** before proceeding to the next section. Press the [**Esc**] key once to exit any command.

Delete Existing Constraints

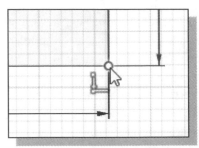

1. Select the bottom **right** corner of the triangle as shown.

   • Note the constraint feasible of the current selection is shown in the *Sketch Palette*; non-feasible constraints are greyed-out.

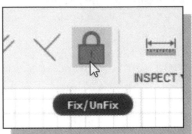

2. Select **Fix/UnFix** to remove the Fix constraint that is applied to the lower right corner of the triangle.

   ❖ Note the removal of the **Fix** constraint has caused the need for **two additional dimensions** to fully define the 2D sketch.

3. On your own, drag the top corner of the triangle upward and note that the entire triangle is free to move in all directions. Release the mouse button to move the triangle to the new location.

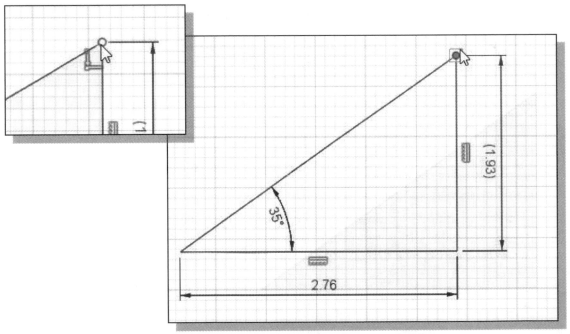

4. On your own, delete the reference dimension on the right and experiment with dragging the corners and the line segments to move the triangle to new locations.

❖ **Dimensional constraints** are used to describe the SIZE and LOCATION of individual geometric shapes. **Geometric constraints** are **geometric restrictions** that can be applied to geometric entities. The constraints applied to the triangle are sufficient to maintain its size and shape, but the geometry can be moved around; its location definition isn't complete.

Add the Location Dimensions

In Autodesk Fusion 360, the Auto Dimension command can be used to assist in creating a fully constrained sketch. **Fully constrained** sketches can be updated more predictably as design changes are implemented. The preferred procedure for fully constrained sketches is to first apply necessary geometric constraints and then use the Sketch Dimension command to add the necessary dimensions. It is also important to realize that different sets of dimensions and geometric constraints can be applied to the same sketch to accomplish a fully constrained geometry.

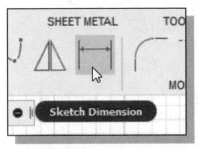

1. Activate the **Sketch Dimension** command in the *Create Panel*.

2. On your own, create the two locational dimensions, measuring the lower right corner to the origin as shown.

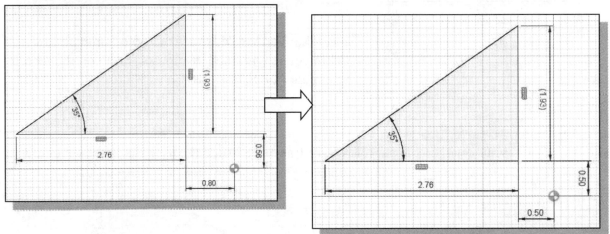

3. On your own, set the two dimensions to **0.5** and observe the alignment of the lower right corner of the triangle to the *Origin*.

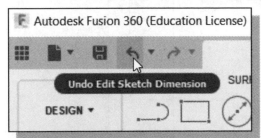

- What is the color of the sketched line?

4. On your own, click the **Undo** button several times to return to the point before the two location dimensions were added.

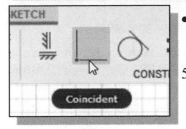

- In Fusion 360, black color indicates fully constrained geometry.

5. Activate the **Coincident Constraint** command in the *Constraints Panel*.

6. Select the **lower right corner** of the triangle and the **Origin** to fully constrain the sketch as shown.

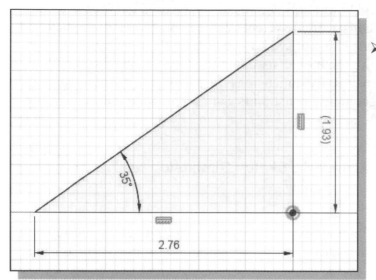

> ➤ In parametric modeling, it is desired to always have fully constrained sketches.

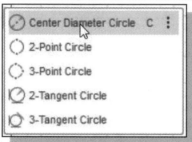

7. Select the **Center Diameter Circle** command by clicking once with the left-mouse-button on the icon in the *Create Panel*.

8. On your own, create a circle of arbitrary size inside the triangle as shown below.

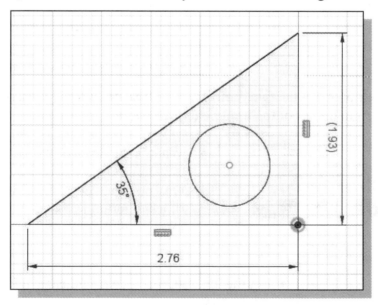

❖ The sketch is not fully constrained with the newly added circle. Which dimensions and/or constraints can be applied to fully constrain the circle? Consider the definitions needed to define the Size and Location of the circle.

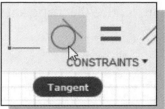

9. Activate the **Tangent** constraint icon in the *Constraints Panel*.

10. Pick the **circle** by left-clicking once on the geometry.

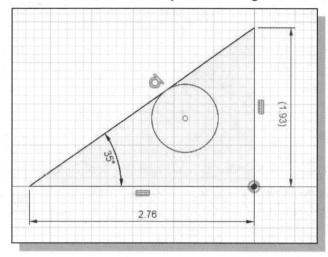

11. Pick the **inclined line**. The sketched geometry is adjusted as shown.

12. Inside the graphics window, click once with the right-mouse-button to display the option menu. Select **OK** in the pop-up menu to end the Tangent command.

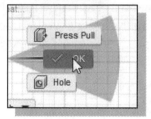

- How many more constraints or dimensions do you think will be necessary to fully constrain the circle? Which constraints or dimensions would you use to fully constrain the geometry?

13. Move the cursor on top of the right side of the circle, and then drag the circle toward the right edge of the graphics window. Notice the size of the circle is adjusted while the system maintains the Tangent constraint.

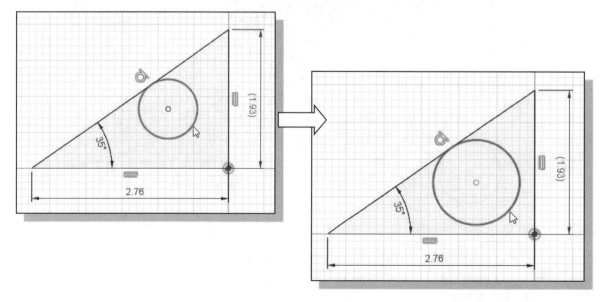

14. Drag the center of the circle toward the upper right direction. Notice the Tangent constraint is always maintained by the system.

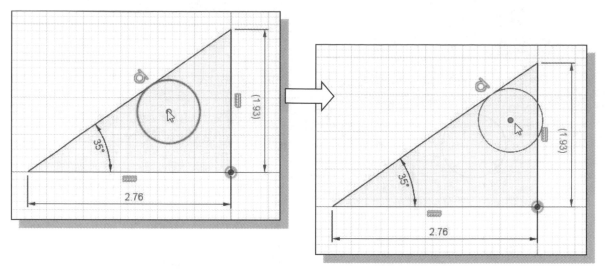

➢ On your own, experiment with adding additional constraints and/or dimensions to fully constrain the sketched geometry. Use the **Undo** command to undo any changes before proceeding to the next section.

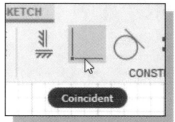

15. Activate the **Coincident Constraint** command in the *Constraints Panel*.

16. Pick the **vertical line**.

17. Pick the **center of the circle** to align the **center** of the circle and the vertical line.

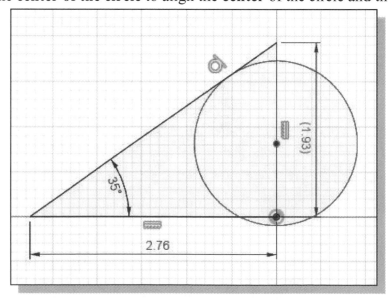

• Is the sketch fully constrained? Can the circle be adjusted?

18. On your own, **delete** the Coincident constraint we just applied. (Hint: click on the coincident icon and hit the [Delete] key.)

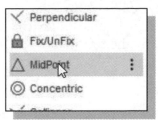

19. On your own, add a Midpoint constraint between the center of the circle and the horizontal line.

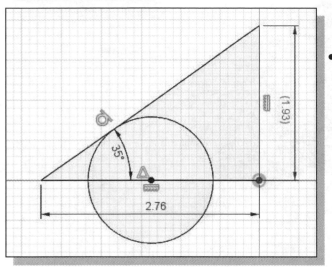

- Is the current sketch fully constrained?

❖ The application of different constraints affects the geometry differently. The design intent is maintained in the CAD model's database and thus allows us to create very intelligent CAD models that can be modified and revised fairly easily. On your own, experiment and observe the results of applying different constraints to the triangle. For example: (1) adding another Fix constraint to the top corner of the triangle; (2) deleting the horizontal dimension and adding another Fix constraint to the left corner of the triangle; and (3) adding another Tangent constraint and adding the size dimension to the circle.

20. On your own, modify the 2D sketch as shown below.

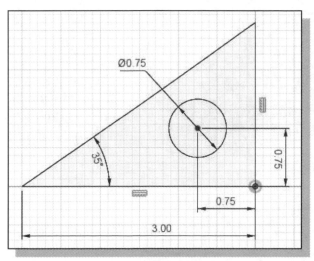

➢ On your own, use the **Extrude** command and create a 3D solid model with a plate thickness of **0.25**. Also, experiment with modifying the dimensions through the *browser* and the *Timeline control*.

Parametric Relations

In parametric modeling, dimensions are design parameters that are used to control the sizes and locations of geometric features. Dimensions are more than just values; they can also be used as feature control variables. This concept is illustrated by the following example.

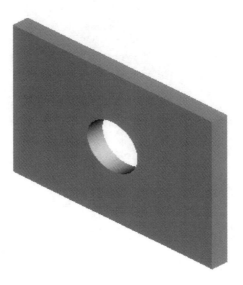

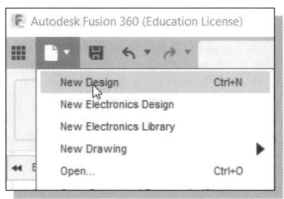

1. Start a new design by left-clicking once on the **File→New Design** icon in the *Standard* toolbar.

   - Another graphics window appears on the screen. We can switch between the two models by clicking on the different tabs.

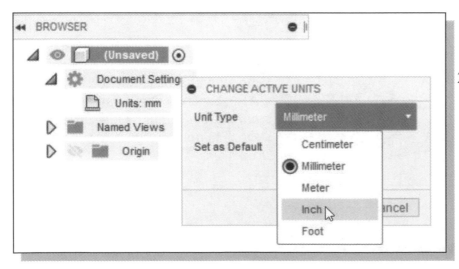

2. On your own, set the document units to **inch** as shown.

3. Select the **Create Sketch** command with the left-mouse-button.

4. Select the *Front XY Plane* as the sketch plane for the new sketch with the **left-mouse-button**.

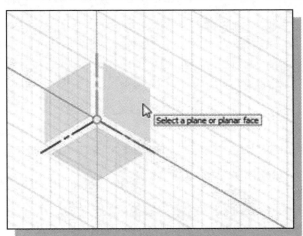

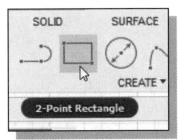

5. Select the **Two point rectangle** command by clicking once with the left-mouse-button on the icon in the *Create Panel*.

6. Create a rectangle of arbitrary size positioned near the center of the screen.

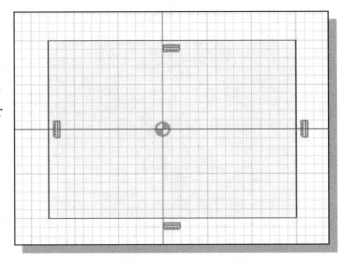

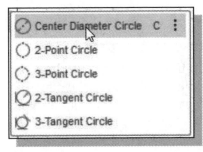

7. Select the **Center Diameter Circle** command by clicking once with the left-mouse-button on the icon in the *Sketch Panel*.

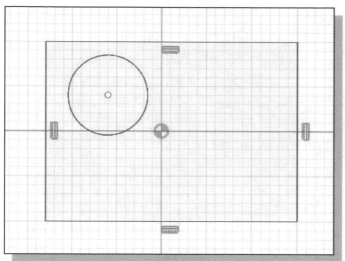

8. Create a **circle** of arbitrary size inside the rectangle as shown.

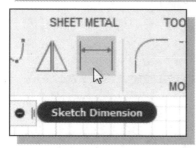

9. Select the **Sketch Dimension** command in the *Sketch Panel*.

10. On your own, create and adjust the geometry by modifying the dimensions (create the overall dimensions, 5.0X3.0 first) as shown below. (Note: Also align the center of the circle to the origin to fully constrain the sketch.)

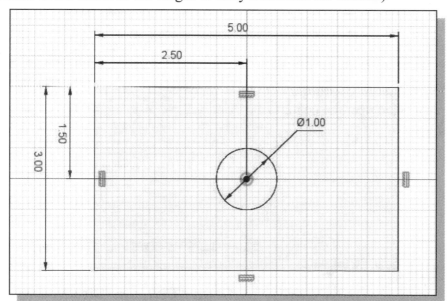

- On your own, change the overall width of the rectangle to **6.0** and the overall height of the rectangle to **3.6** and observe the location of the circle in relation to the edges of the rectangle. Adjust the dimensions back to **5.0** and **3.0** as shown in the above figure before continuing.

Dimensional Values and Dimensional Variables

Initially in Autodesk Fusion 360, values are used to create different geometric entities. The text created by the **Dimension** command also reflects the actual location or size of the entity. Each dimension is also assigned a name that allows the dimension to be used as a control variable. The default format is "dxx," where the "xx" is a number that Autodesk Fusion 360 increments automatically each time a new dimension is added.

Let us look at our current design, which represents a plate with a hole at the center. The dimensional values describe the size and/or location of the plate and the hole. If a modification is required to change the width of the plate, the location of the hole will remain the same as described by the two location dimensional values. This is okay if that is the design intent. On the other hand, the *design intent* may require (1) keeping the hole at the center of the plate and (2) maintaining the size of the hole to be one-third of the height of the plate. We will establish a set of parametric relations using the dimensional variables to capture the design intent described in statements (1) and (2) above.

1. Move the cursor on the **width dimension** of the rectangle and notice the dimension variable name, **d1**, is displayed.

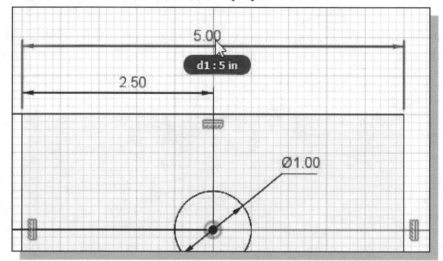

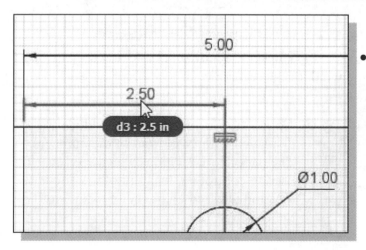

*   Notice each dimension has an associated *variable name* **dx**. The variable names are automatically created with the Sketch Dimension command.

Parametric Equations

Each time we add a dimension to a model, that value is established as a parameter for the model. We can use parameters in equations to set the values of other parameters.

1. Double-click on the **horizontal location dimension**, 2.5, of the circle to display the *Edit Dimension* window.

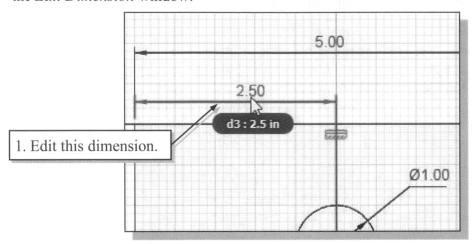

2. Click on the width dimension of the rectangle **d1** (value of **5.0**). Notice the selected variable name is automatically entered in the *Edit Dimension* window.

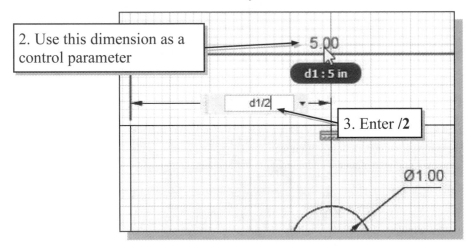

3. In the *Edit Dimension* window, enter **/2** to set the horizontal location dimension of the circle to be one-half of the width of the rectangle.

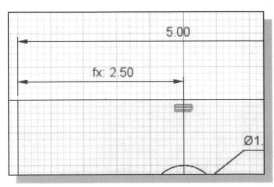

❖ Notice the derived dimension values are displayed with **fx** in front of the numbers. The parametric relations we entered are used to control the location of the circle; the location is based on the height and width of the rectangle.

4. On your own, repeat the above steps and set the vertical location dimension to one-half of the height of the rectangle.

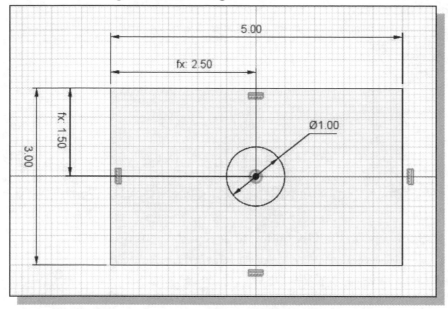

Viewing the Established Parameters and Relations

1. In the *Modify* tool panel select the **Change Parameters** command by left-clicking once on the icon.

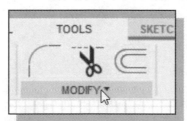   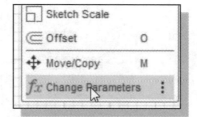

*   The **Change Parameters** command can be used to display all dimensions used to define the model. We can also create additional parameters as design variables, which are called ***user parameters***. Expand the sketch1 parameters as shown.

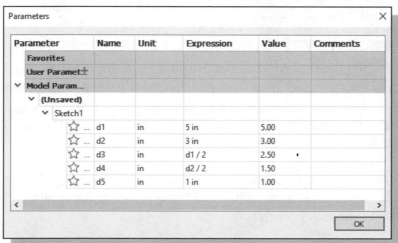

Parameter	Name	Unit	Expression	Value	Comments
Favorites					
User Paramet⁚					
⌄ **Model Param...**					
⌄ **(Unsaved)**					
⌄ Sketch1					
☆ ...	d1	in	5 in	5.00	
☆ ...	d2	in	3 in	3.00	
☆ ...	d3	in	d1 / 2	2.50	•
☆ ...	d4	in	d2 / 2	1.50	
☆ ...	d5	in	1 in	1.00	

OK

2. Click on the **1.0** value in the equation section of the *Parameters* window. Enter **d2/3** as the parametric relation to set the size of the circle to be one-third of the height of the rectangle as shown below.

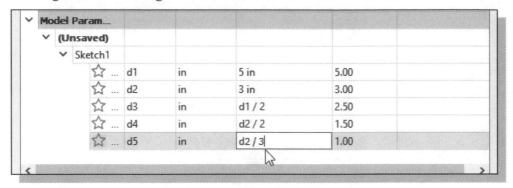

     3. Click on the **OK** button to accept the settings.

4. On your own, change the dimensions of the rectangle to **6.0 x 3.6** and observe the changes to the location and size of the circle. (Hint: Double-click the dimension text to enter the *Edit Dimension* mode.)

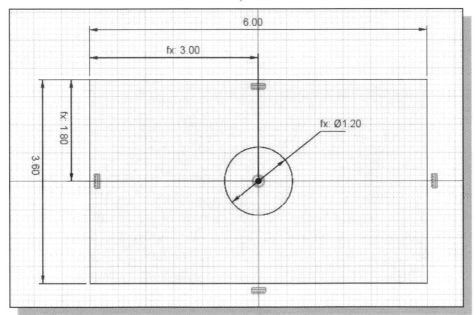

❖ *Autodesk Fusion 360* automatically adjusts the dimensions of the design, and the parametric relations we entered are also applied and maintained. The dimensional constraints are used to control the size and location of the hole. The design intent, previously expressed by statements (1) and (2) at the beginning of this section, is now embedded into the model.

➢ On your own, use the **Extrude** command and create a 3D solid model with a plate thickness of **0.25**. Also, experiment with modifying the parametric relations and dimensions through the *part browser*.

Use the Measure Tools

Besides using the measure tools to get geometric information at the 2D level, the measure tools can also be used on 3D models.

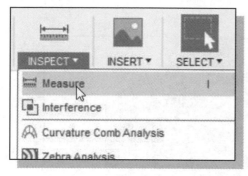

1. In the *Inspect Panel*, left-click once on the **Measure** option as shown.

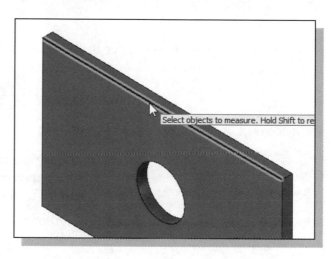

2. Click on the **top edge** of the rectangular plate as shown.

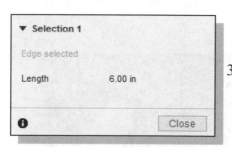

3. The associated length measurement of the selected geometry is displayed in the *Measure* dialog box as shown.

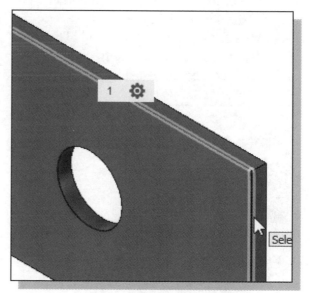

4. Click on the **right edge** of the rectangular plate as shown. The associated information regarding the selected entities is shown.

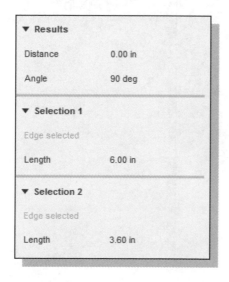

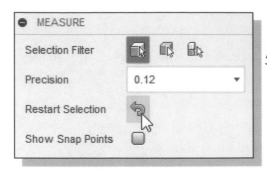

5. Click **Restart Selection** as shown.

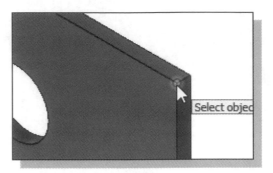

6. Click the **top-right corner** of the 3D model as shown.

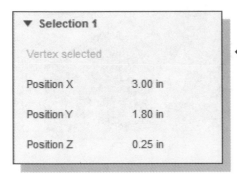

❖ Notice the information regarding the selected point is displayed in the *Measure Distance* box. The absolute position of the point is displayed. (Note the displayed numbers may be different on your screen.)

7. Select the front left bottom corner as the second location for the **Measure Length** command. The distance in between the two selected objects is calculated.

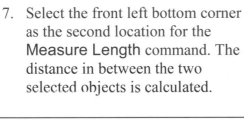

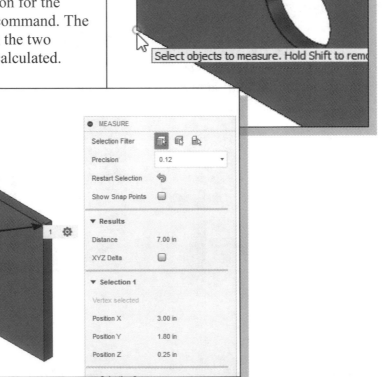

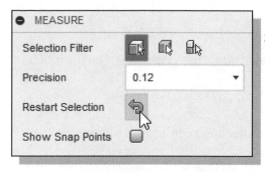

8. Click **Restart Selection** as shown.

9. Click on the front face of the plate model to display the area of the selected surface.

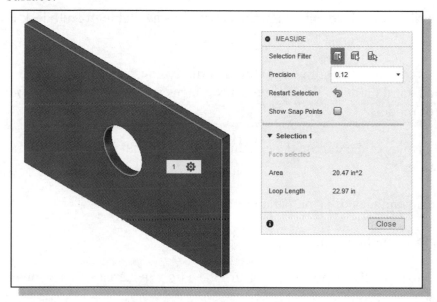

10. Note that we can also select cylindrical surfaces; select the cylindrical surface as shown.

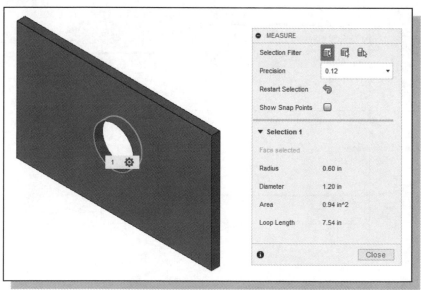

11. On your own, experiment with the different **measure tools** available.

Review Questions:

1. What is the difference between *dimensional* constraints and *geometric* constraints?

2. How can we confirm that a sketch is fully constrained?

3. How do we distinguish between derived dimensions and regular dimensions on the screen?

4. Describe the procedure to Display/Edit user-defined equations.

5. List and describe three different geometric constraints available in Autodesk Fusion 360.

6. Does Autodesk Fusion 360 allow us to build partially constrained or totally unconstrained solid models? What are the advantages and disadvantages of building these types of models?

7. How do we display and examine the existing constraints that are applied to the sketched entities?

8. Describe the advantages of using parametric equations.

9. Can we delete an applied constraint? How?

10. Create the following 2D Sketch and measure the associated area and perimeter.

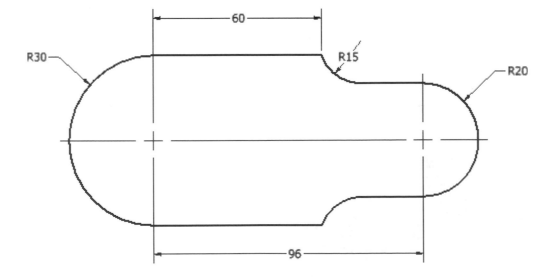

Exercises: Create and save the exercises in the Chapter5 folder.
(Create and establish three parametric relations for each of the following designs.)

1. **Swivel Base** (Dimensions are in millimeters. Base thickness: **10 mm.** Boss: **5 mm.**)

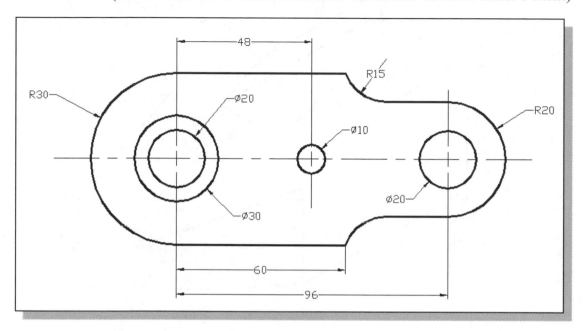

2. **Anchor Base** (Dimensions are in inches.)

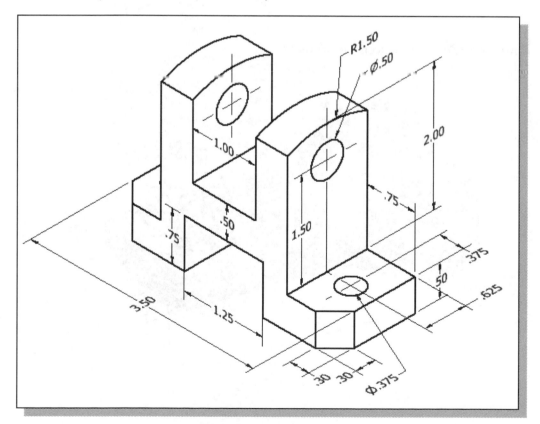

3. **Wedge Block** (Dimensions are in inches.)

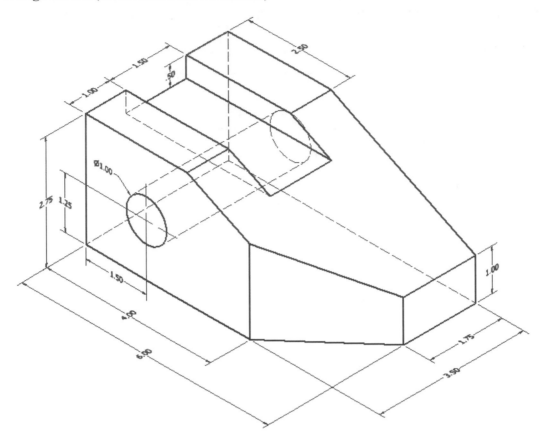

4. **Hinge Guide** (Dimensions are in inches.)

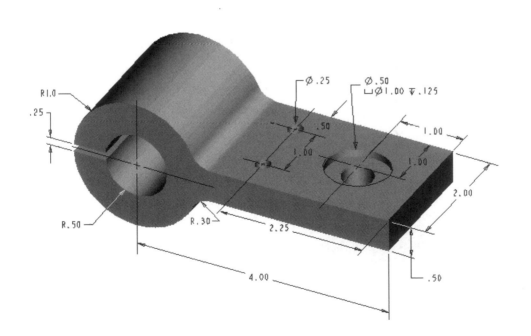

5. **Pivot Holder** (Dimensions are in inches.)

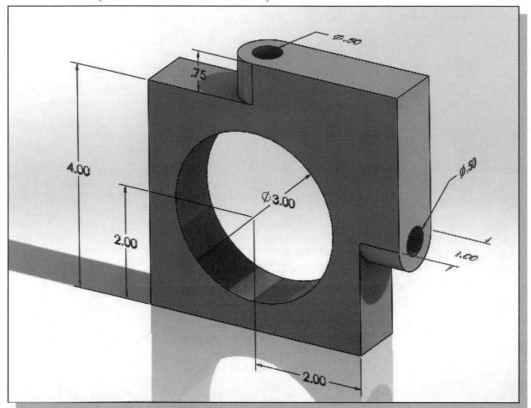

6. **Support Fixture** (Dimensions are in inches.)

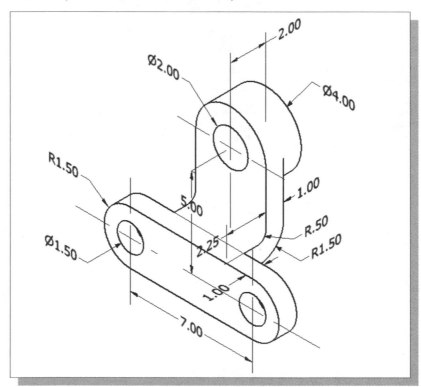

Notes:

Chapter 6
Geometric Construction Tools

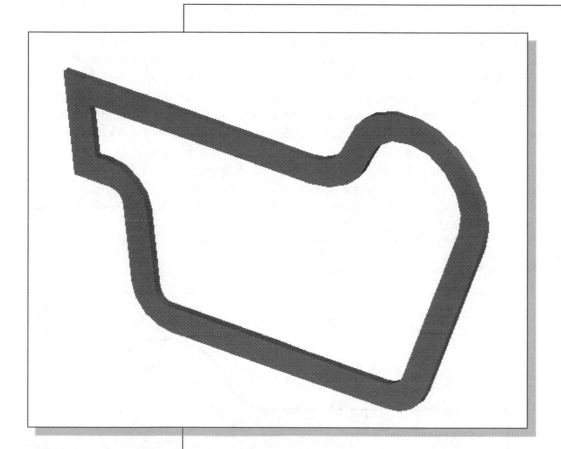

Learning Objectives

- ♦ **Apply Geometry Constraints**
- ♦ **Use the Trim/Extend Command**
- ♦ **Use the Offset Command**
- ♦ **Understand the Profile Sketch Approach**
- ♦ **Create Projected Geometry**
- ♦ **Understand and Use Reference Geometry**
- ♦ **Edit with Click and Drag**
- ♦ **Use the Auto Dimension Command**

Introduction

The main characteristics of solid modeling are the accuracy and completeness of the geometric database of the three-dimensional objects. However, working in three-dimensional space using input and output devices that are largely two-dimensional in nature is potentially tedious and confusing. Autodesk Fusion 360 provides an assortment of two-dimensional construction tools to make the creation of wireframe geometry easier and more efficient. Autodesk Fusion 360 includes two types of wireframe geometry: *curves* and *profiles*. Curves are basic geometric entities such as lines, arcs, etc. Profiles are a group of curves used to define a boundary. A *profile* is a closed region and can contain other closed regions. Profiles are commonly used to create extruded and revolved features. An *invalid profile* consists of self-intersecting curves or open regions. In this lesson, the basic geometric construction tools, such as Trim and Extend, are used to create profiles. The Autodesk Fusion 360's *profile sketch* approach to creating profiles is also introduced. Mastering the geometric construction tools along with the application of proper constraints and parametric relations is the true essence of *parametric modeling*.

The Gasket Design

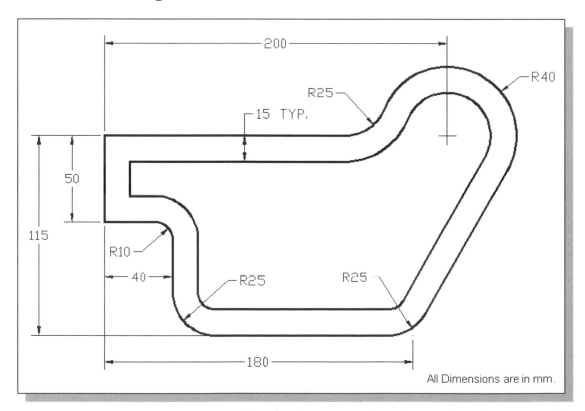

❖ Based on your knowledge of Autodesk Fusion 360 so far, how would you create this design? What is the most difficult geometry involved in the design? Take a few minutes to consider a modeling strategy and do preliminary planning by sketching on a piece of paper. You are also encouraged to create the design on your own prior to following through the tutorial.

Modeling Strategy

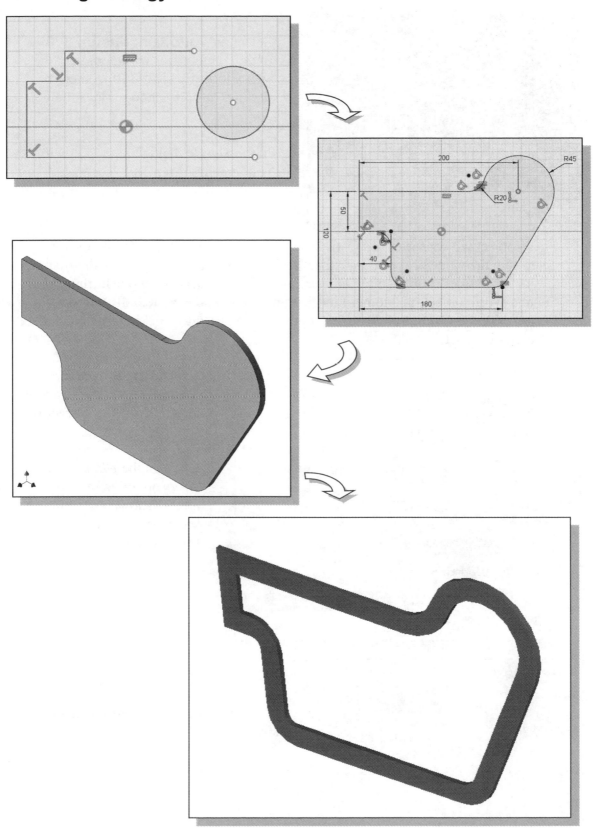

Starting Autodesk Fusion 360

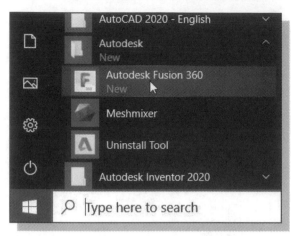

1. Select the **Autodesk Fusion 360** option on the *Start* menu or select the **Autodesk Fusion 360** icon on the desktop to start Autodesk Fusion 360.

2. In the *Sign In* dialog box, log in with your email or username.

❖ Every object we construct in a CAD system is measured in units. We should determine the value of the units within the CAD system before creating the first geometric entities. For example, in one model, a unit might equal one millimeter of the real-world object; in another model, a unit might equal an inch. In Autodesk Fusion 360, the *Change Active Units* option can be used to quickly switch to using a different units set.

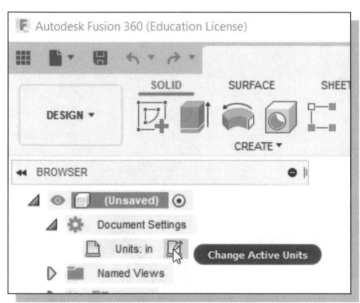

3. Expand the *Document Setting* item in the Browser area to show the current units setup as shown.

4. Click on the *Change Active Units* option in the Browser area as shown. We will use the millimeter (mm) units for this example.

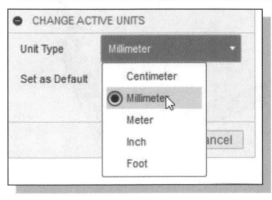

5. In the *Units* list, select **Millimeter** as shown.

6. Click on the **OK** button to accept the new setting.

Create the first 2D Sketch

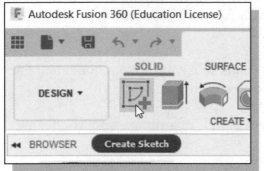

1. In the *Create toolbar* select the **Create Sketch** command by left-clicking once on the icon.

2. Move the cursor over the right-vertical plane (the *XY Plane)* in the graphics area. Click once with the **left-mouse-button** to select the *Plane* as the sketch plane for the new sketch.

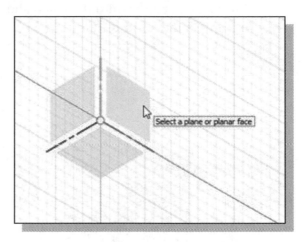

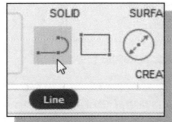

3. Click on the **Line** icon in the *Sketch panel* on the Ribbon.

4. Create a sketch as shown in the figure below. Start the sketch from the top right corner. The line segments are all parallel and/or perpendicular to each other. We will intentionally make the line segments of arbitrary length, as it is quite common during the design stage that not all of the shapes and forms are determined.

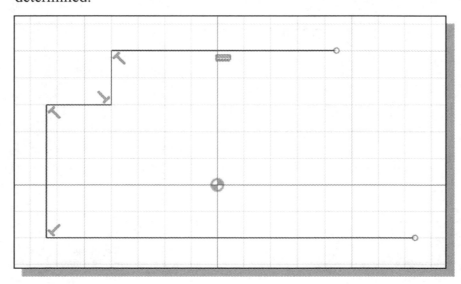

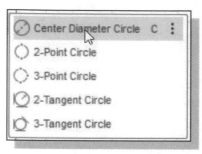

5. Select the **Center Diameter Circle** command by clicking once with the left-mouse-button on the icon in the *Sketch Panel*.

6. Pick a location that is above the bottom horizontal line as the center location of the circle.

7. Move the cursor toward the right, and create a circle of arbitrary size by clicking once with the left-mouse-button.

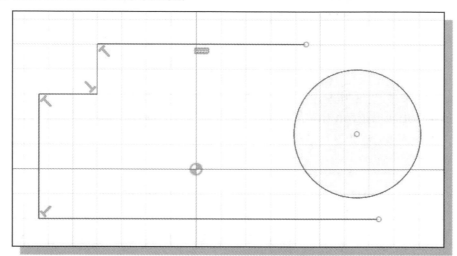

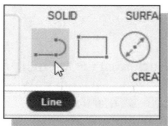

8. Click on the **Line** icon in the *Sketch panel* on the Ribbon.

9. Move the cursor near the right end point on the bottom horizontal line as shown.

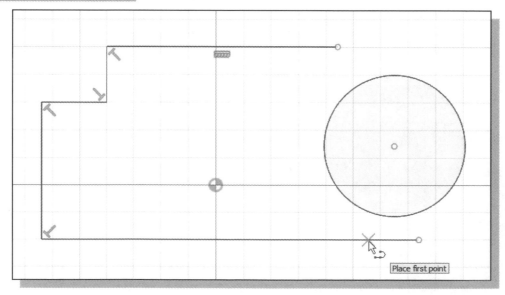

10. For the other end of the line, select a location that is on the upper right side of the circle as shown.

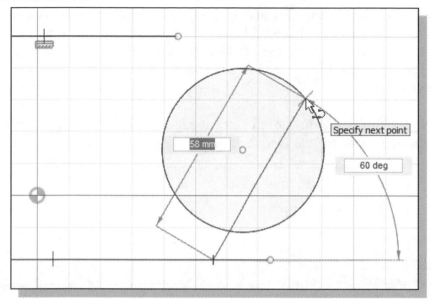

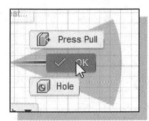

11. Inside the graphics window, right-click to bring up the option menu and select **OK** to end the Line command.

- Notice the two coincident constraints applied to the two endpoints of the line we just created. The coincident constraints indicate the endpoints are always on the circle and the bottom horizontal line.

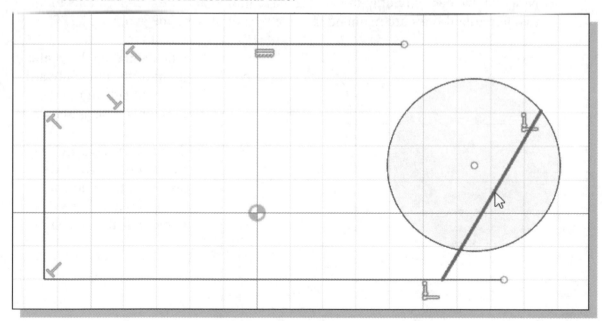

Edit the Sketch by Dragging the Sketched Entities

In Autodesk Fusion 360, we can click and drag any under-constrained curve or point in the sketch to change the size or shape of the sketched profile. As illustrated in the previous chapter, this option can be used to identify under-constrained entities. This *Editing by Dragging* method is also an effective visual approach that allows designers to quickly make changes.

1. Move the cursor on the lower left vertical edge of the sketch. Click and drag the edge to a new location that is toward the right side of the sketch.

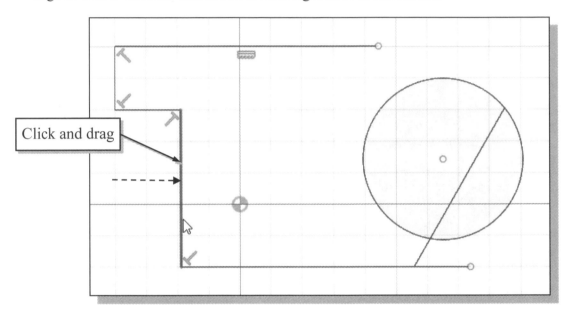

Click and drag

❖ Note that we can only drag the vertical edge horizontally; the connections to the two horizontal lines are maintained while we are moving the geometry.

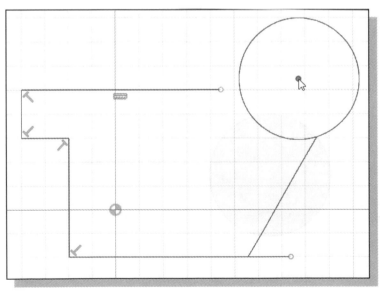

2. Click and drag the center point of the circle toward the right side of the top horizontal line as shown.

❖ The *Editing by dragging* method is an effective approach that allows designers to quickly explore different design concepts.

Adding Dimensions and Constraints

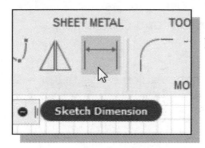

1. Activate the **Sketch Dimension** command in the *Sketch Panel*.

2. Add the horizontal location dimension, from the top left vertical edge to the center of the circle as shown. (Do not be overly concerned with the dimensional value; we are still working on creating a *rough sketch*.)

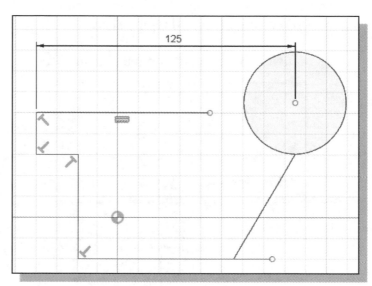

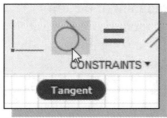

3. Click on the **Tangent** constraint icon in the *Sketch Palette* panel.

4. Pick the inclined line by left-clicking once on the geometry.

5. Pick the circle as the second entity to the tangent constraint.

6. The sketched geometry is adjusted as shown below.

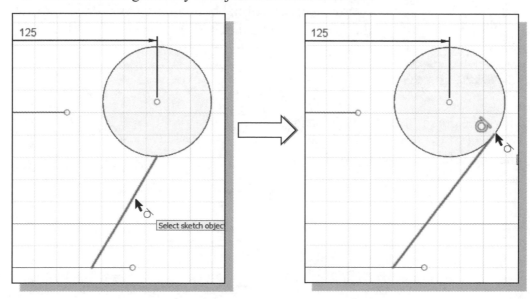

Use the Trim and Extend Commands

In the following sections, we will illustrate using the Trim and Extend commands to complete the desired 2D profile.

The **Trim** and **Extend** commands can be used to shorten/lengthen an object so that it ends precisely at a boundary. As a general rule, Autodesk Fusion 360 will try to clean up sketches by forming a closed region sketch.

1. Activate the **Extend** command in the *Sketch* panel as shown.

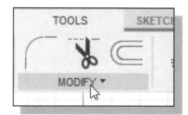 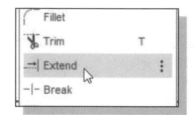

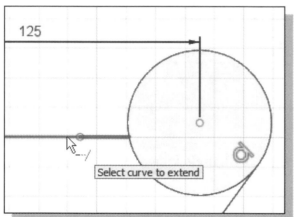

2. We will first extend the top horizontal line to the circle. Move the cursor near the right hand endpoint of the top horizontal line. Autodesk Fusion 360 will automatically display the possible result of the selection.

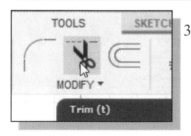

3. Next, we will use the **Trim** command to remove the overhang portion of the bottom horizontal line. Activate the Trim command in the *Sketch* panel.

4. Move the cursor near the right hand endpoint of the bottom horizontal line and select the line segment to trim.

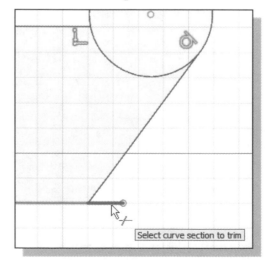

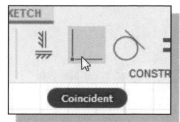

5. On your own, apply a **coincident constraint** on the top horizontal line and the center point of the circle as shown below.

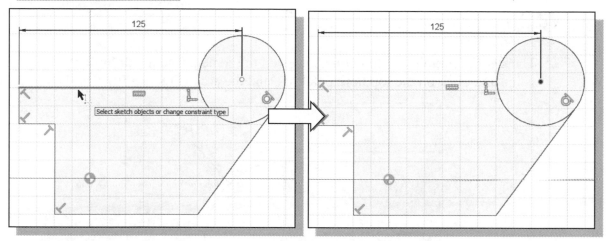

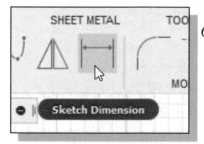

6. On your own, **trim** the circle and add the additional **dimensions** as shown. (At this point, accept the default values of all dimensions; the dimension values will be adjusted in the next sections.)

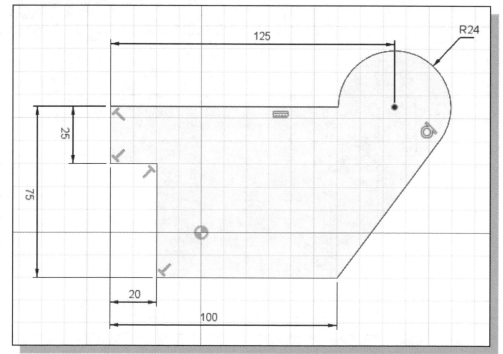

Create Fillets and Completing the Sketch

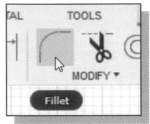

1. Activate the **Fillet** command in the *Sketch* panel as shown.

2. The *2D Fillet* radius dialog box appears on the screen. Use the **default** radius value and click on the top horizontal line and the arc to create a fillet as shown.

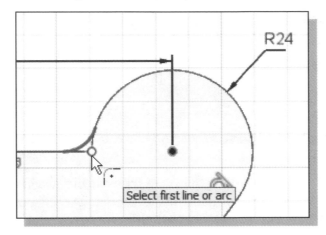

3. On your own, create the three additional fillets as shown in the below figure. Note that the **Equal** constraint is activated, and all rounds and fillets are created with the constraint.

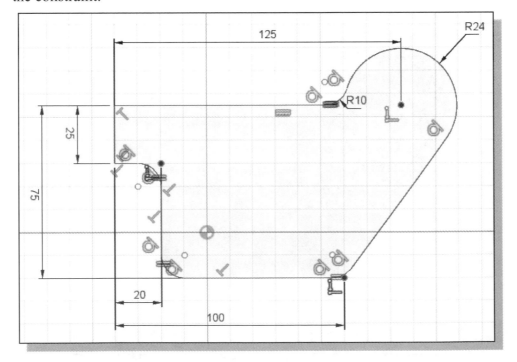

Fully Constrained Geometry

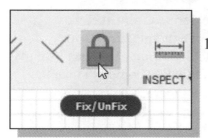

1. Click on the **Fix** constraint icon in the *Sketch Palette Panel*.

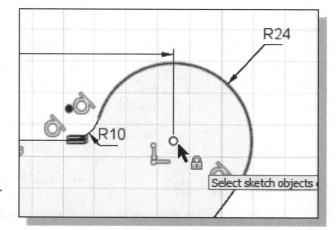

2. Apply the **Fix** constraint to the center of the arc as shown below.

➢ Note that the **Black** color indicates the sketch is now fully constrained.

3. On your own, complete the sketch by adjusting the dimensions to the desired values as shown below. (Hint: Change the overall width/height and the radius of fillets first.)

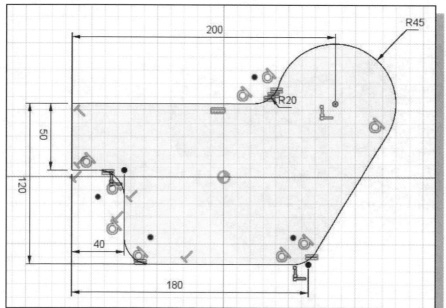

❖ Note that by applying proper geometric and dimensional constraints to the sketched geometry; a *constraint network* is created that assures the geometry shape behaves predictably as changes are made.

4. Click on the **Stop Sketch** icon to exit the 2D sketch mode.

Profile Sketch

In Autodesk Fusion 360, *profiles* are closed regions that are defined from sketches. Profiles are used as cross sections to create solid features. For example, **Extrude**, **Revolve**, **Sweep**, **Loft**, and **Coil** operations all require the definition of at least a single profile. The sketches used to define a profile can contain additional geometry since the additional geometry entities are consumed when the feature is created. To create a profile we can create single or multiple closed regions, or we can select existing solid edges to form closed regions. A profile cannot contain self-intersecting geometry; regions selected in a single operation form a single profile. As a general rule, we should dimension and constrain profiles to prevent them from unpredictable size and shape changes. Autodesk Fusion 360 does allow us to create under-constrained or non-constrained profiles; the dimensions and/or constraints can be added/edited later.

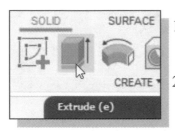

1. In the *Create panel*, select the **Extrude** command by left-clicking once on the icon.

2. Select the inside region of the 2D sketch as the profile for the extrude command.

3. In the *Extrude* dialog box, enter **5 mm** as the extrusion distance as shown.

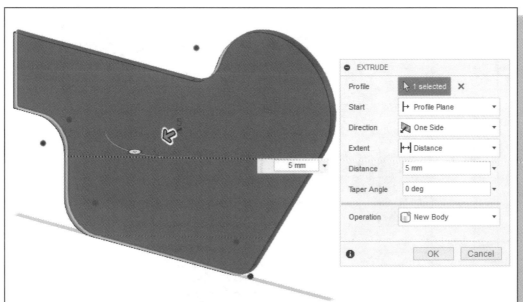

4. Click on the **OK** button to accept the settings and create the solid feature.

- Note that Autodesk Fusion 360 uses the selected closed region of the sketch to form the **profile** required for the *Extrude* operation.

Redefine the Sketch and Profile

Engineering designs usually go through many revisions and changes. Autodesk Fusion 360 provides an assortment of tools to handle design changes quickly and effectively. We will demonstrate some of the tools available by changing the base feature of the design. The profile used to create the extrusion is selected from the sketched geometry entities. In Autodesk Fusion 360, any profile can be edited and/or redefined at any time. It is this type of functionality in parametric solid modeling software that provides designers with greater flexibility and the ease to experiment with different design considerations.

1. In the *browser*, **right-click** on **Sketch1** feature to show the option menu and select the **Edit Sketch** option as show.

2. On your own, create a circle and a rectangle of arbitrary sizes, positioned as shown in the figure below. We will intentionally under-constrain the new sketch to illustrate the flexibility of the system.

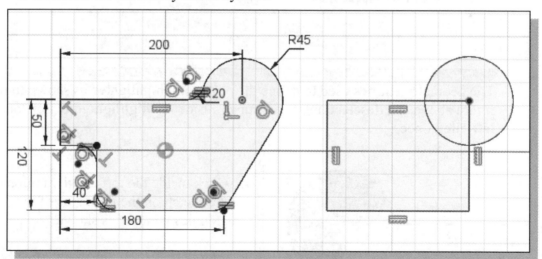

3. Select **Stop Sketch** in the *Ribbon toolbar* to end the Sketch option.

➤ Note that, at this point, the solid model remains the same as before. We will next redefine the profile used for the base feature.

4. In the *Timeline control panel*, double-click on the Extrude1 feature to enter the **Edit Feature** mode.

❖ Autodesk Fusion 360 will now display the feature elements in the feature control panel inside the graphics window. We have literally gone back in time to the point where we defined the extrusion feature.

5. In the *Edit Feature* control panel, the currently selected **Profile** is highlighted. Fusion 360 indicates one region is selected for the profile.

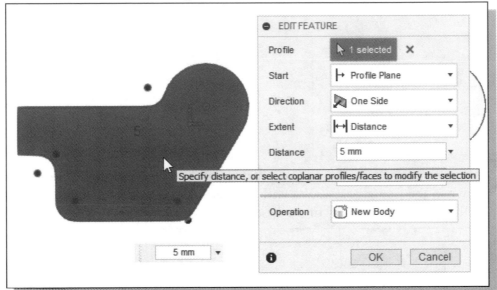

6. The geometry entities used to define the profile are highlighted as shown in the figure above. To **deselect**, move the cursor inside the **highlighted** region, and **left-click** once.

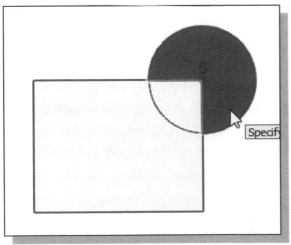

7. Click inside the top circular region of the sketch on the right side of the graphics window as shown.

➢ Autodesk Fusion 360 automatically selects the geometry entities that form a closed region to define the profile.

❖ The original sketch and the newly added entities are wireframe entities recorded by the system as belonging to the same **SKETCH**, but only the **PROFILED** entities are used to create the feature.

8. Click inside the circle to complete the profile definition as shown in the figure below.

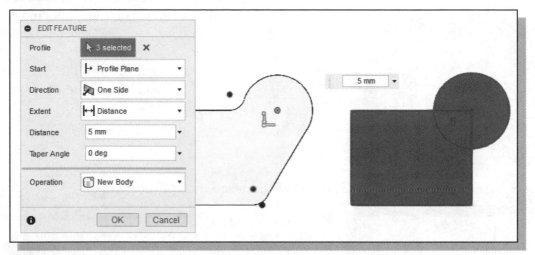

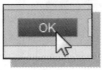

9. In the *Extrude* control panel, click on the **OK** button to accept the settings and update the solid feature.

• The feature is recreated using the newly sketched geometric entities, which are under-constrained and with no dimensions. The profile is created with extra wireframe entities by selecting multiple regions. The extra geometry entities can be used as construction geometry to help in defining the profile. This approach encourages engineering content over drafting technique, which is one of the key features of Autodesk Fusion 360 over other solid modeling software.

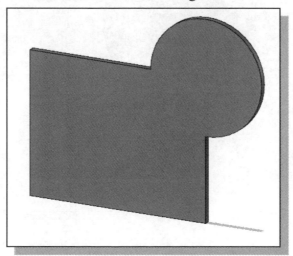

10. On your own, repeat the above steps and **reset** the profile back to the original gasket sketch as shown on page 14.

Create an Offset Cut Feature

To complete the design, we will create a cutout feature by using the **Offset** command. First, we will set up the sketching plane to align with the front face of the 3D model.

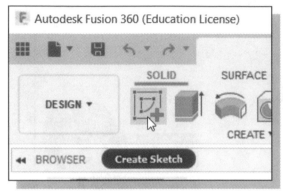

1. In the *Sketch* panel select the **Create Sketch** command by left-clicking once on the icon.

2. Select the **Front face** of the 3D model in the graphics window to align the sketching plane of the new sketch.

3. Activate the **Project Geometry** command in the *Create* panel.

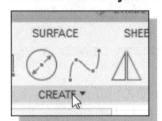
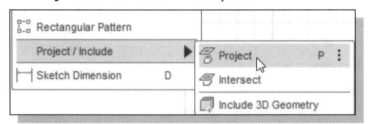

4. Select the front face of the gasket model and notice the outline of the design has been projected onto the sketching plane.

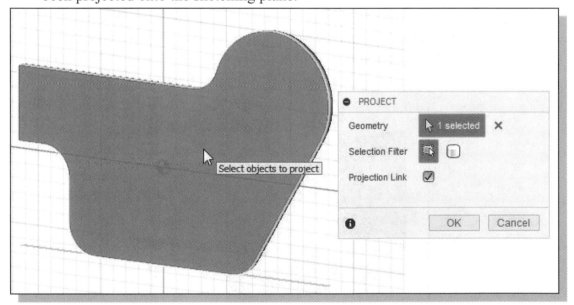

5. Click on the **OK** button to accept the settings and exit the **Project Geometry** command.

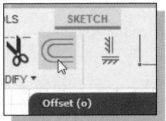

6. Activate the **Offset** command in the *Sketch panel*.

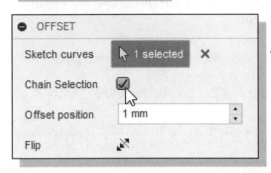

7. In the *Offset* dialog box, confirm the **Chain** option is activated as shown.

8. Select any edge of the **front face** of the 3D model. Autodesk Fusion 360 will automatically select all of the connecting geometry to form a closed region.

9. In the *graphics window*, drag the direction control downward to **15 mm** as the offset distance as shown.

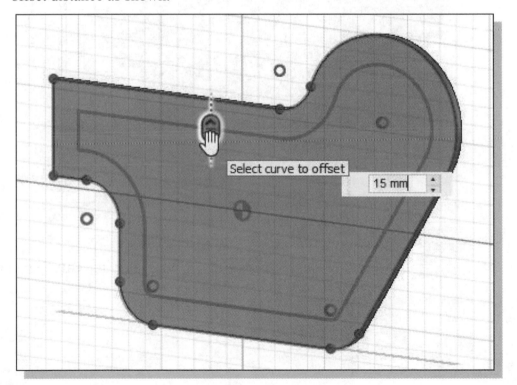

10. Click on the **OK** button to accept the settings and exit the **Project Geometry** command.

11. Select **Stop Sketch** in the toolbar area to end the Sketch option.

12. In the *Create panel,* activate the **Extrude** command by left-clicking once on the icon.

13. Select the **inside region** of the offset geometry as the profile for the extrusion.

14. Drag the **arrow control**, with the left mouse button, to set the extrusion direction toward the back as shown.

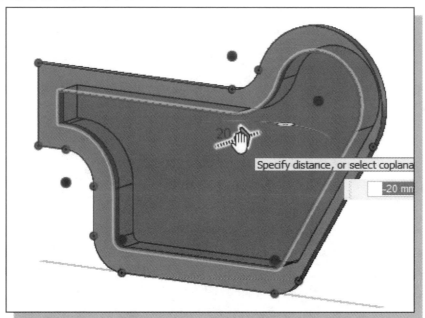

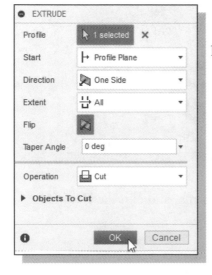

15. In the *Extrude* dialog box, select the **Cut** operation, set the *Extent* to **All** as shown.

16. In the *Extrude* dialog box, click on the **OK** button to accept the settings and create the solid feature.

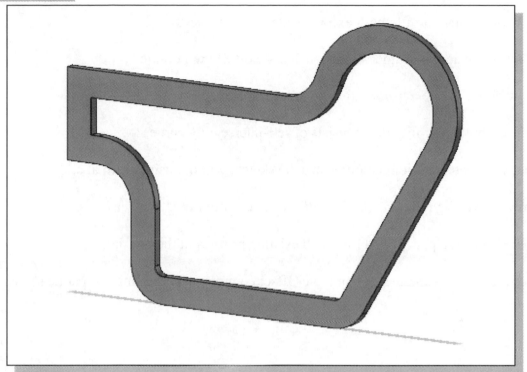

➤ The offset geometry is associated with the original geometry. On your own, adjust the overall height of the design to **150 mm** and offset distance to **18mm**; confirm that the offset geometry and the solid model are adjusted accordingly.

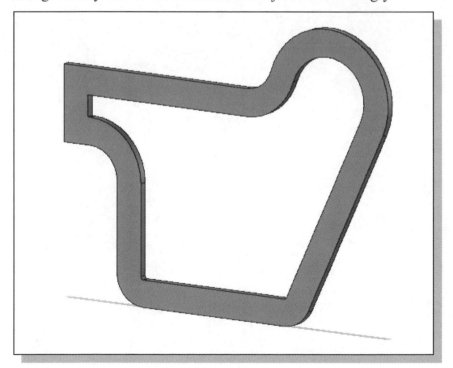

Review Questions:

1. What are the two types of wireframe geometry available in Autodesk Fusion 360?

2. Can we create a profile with extra 2D geometry entities?

3. How do we access Autodesk Fusion 360's **Edit Sketch** option?

4. How do we create a *profile* in Autodesk Fusion 360?

5. Can we build a profile that consists of self-intersecting curves?

6. Can we create additional entities in a 2D sketch, without using them at all?

7. How do we align the *sketch plane* of a selected entity to the screen?

8. Describe the steps we used to switch existing profiles in the tutorial.

9. Describe the advantages of using the Offset command vs. creating a separate sketch.

Exercises: Create and save the exercises in the Chapter 6 folder.

1. **V-slide Plate** (Dimensions are in inches. Plate Thickness: **0.25**)

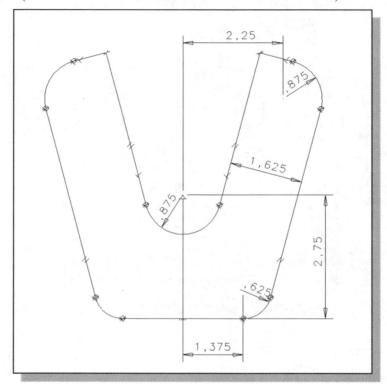

2. **Shaft Support** (Dimensions are in millimeters. Note the two R40 arcs at the base share the same center.)

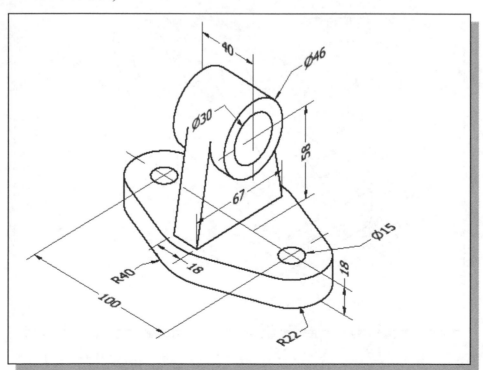

3. **Vent Cover** (Thickness: **0.125** inches. Hint: Use the Ellipse command.)

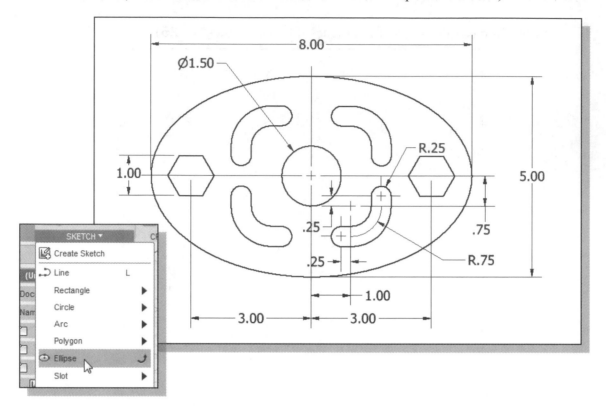

4. **Anchor Base** (Dimensions are in inches.)

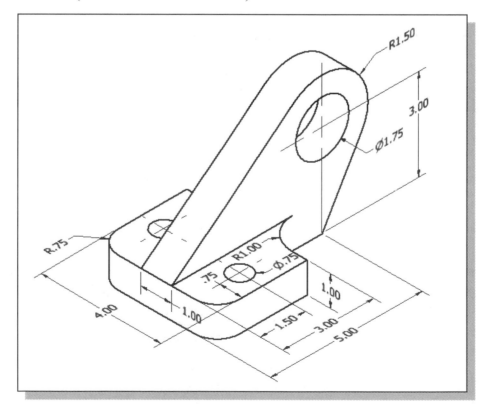

5. **Tube Spacer** (Dimensions are in inches.)

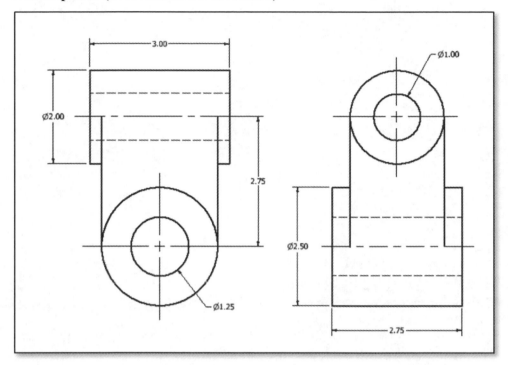

6. **Pivot Lock** (Dimensions are in inches. The circular features in the design are all aligned to the two centers at the base.)

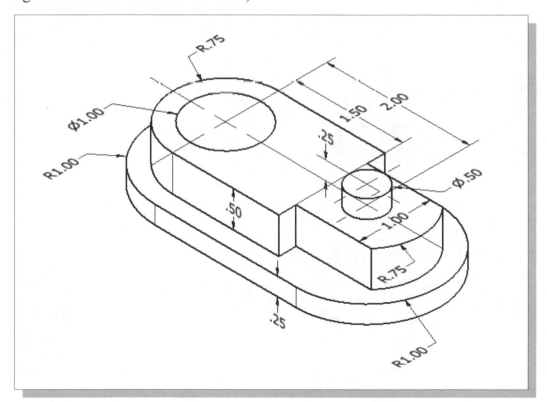

Notes:

Chapter 7
Parent/Child Relationships and the BORN Technique

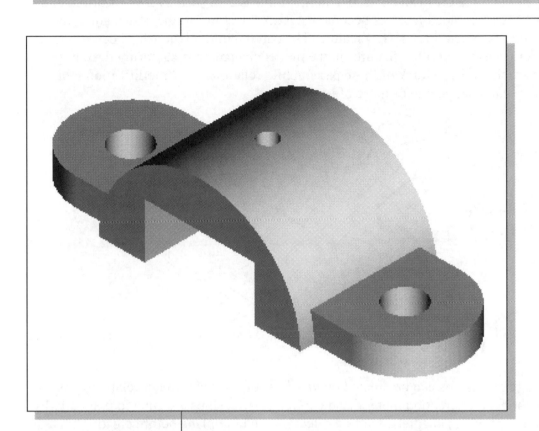

Learning Objectives

♦ **Understand the Concept and Usage of the BORN Technique**
♦ **Understand the Importance of Parent/Child Relations in Features**
♦ **Use the Suppress Feature Option**
♦ **Resolve Undesired Feature Interactions**

Introduction

The parent/child relationship is one of the most powerful aspects of *parametric modeling*. In Autodesk Fusion 360, each time a new modeling event is created, previously defined features can be used to define information such as size, location, and orientation. The referenced features become **PARENT** features to the new feature, and the new feature is called the **CHILD** feature. The parent/child relationships determine how a model reacts when other features in the model change, thus capturing design intent. It is crucial to keep track of these parent/child relations. Any modification to a parent feature can change one or more of its children.

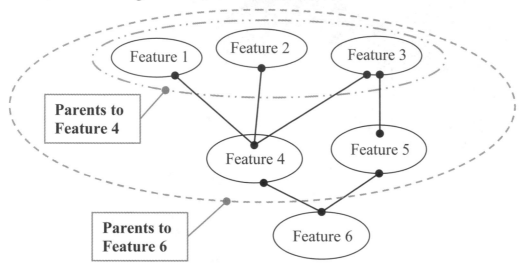

Parent/child relationships can be created *implicitly* or *explicitly*; implicit relationships are implied by the feature creation method and explicit relationships are entered manually by the user. In the previous chapters, we first select a sketching plane before creating a 2D profile. The selected surface becomes a parent of the new feature. If the sketching plane is moved, the child feature will move with it. As one might expect, parent/child relationships can become quite complicated when the features begin to accumulate. It is therefore important to think about modeling strategy before we start to create anything. The main consideration is to try to plan ahead for possible design changes that might occur which would be affected by the existing parent/child relationships. Parametric modeling software, such as Autodesk Fusion 360, also allows us to adjust feature properties so that any feature conflicts can be quickly resolved.

The BORN Technique

In *parametric modeling*, the **base feature** (the *first solid feature* of the solid model) is the center of all features and is considered the key feature of the design. All subsequent features are built by referencing the first feature. Much emphasis is placed on the selection of the *base feature*.

A more advanced technique of creating solid models is what is known as the "**Base Orphan Reference Node**" (**BORN**) technique. The basic concept of the BORN technique is to use a *Cartesian coordinate system* as the first feature prior to creating any solid features. With the *Cartesian coordinate system* established, we then have three mutually perpendicular datum planes (namely the *XY, YZ,* and *ZX planes*), three datum axes and a datum point available to use as sketching planes. The three datum planes can also be used as references for dimensions and geometric constructions. Using this technique, the first node in the history tree is called an "orphan," meaning that it has no history to be replayed. The technique of using the reference geometry in this "base node" is therefore called the BORN technique.

Autodesk Fusion 360 automatically establishes a set of reference geometry when we start a new part, namely a *Cartesian coordinate system* with three work planes, three work axes, and a work point. All subsequent solid features can then use the coordinate system and/or reference geometry as sketching planes. The *base feature* is still important, but the *base feature* is no longer the <u>ONLY</u> choice for creating subsequent solid features. This approach provides us with more options while we are creating parametric solid models. More importantly, this approach provides greater flexibility for part modifications and design changes. This approach is also very useful in creating assembly models, which will be illustrated in the later chapters of this text.

The U-Bracket Design

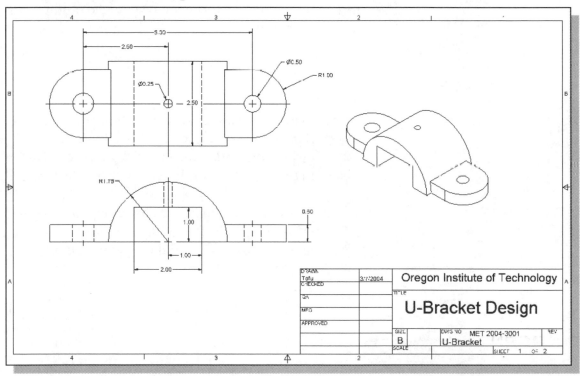

> Based on your knowledge of Fusion 360 so far, how many features would you use to create the model? Which feature would you choose as the **base feature**? What is your choice for arranging the order of the features? Would you organize the features differently if the rectangular cut at the center is changed to a circular shape (Radius: 1.25 inch)?

Apply the BORN Technique

1. Select the **Autodesk Fusion 360** option on the *Start* menu or select the **Autodesk Fusion 360** icon on the desktop to start Autodesk Fusion 360.

2. On your own, set the *document units* to **inch**.

3. In the *Part Browser* window, click on the triangular symbol in front of the ***Origin*** feature to display the pre-defined datum items.

❖ In the *Part Browser* window, notice a new part name appeared with seven datum features established. The seven datum features include three *work planes*, three *work axes*, and a *work point*. By default, the three work planes and work axes are aligned to the **world coordinate system** and the work point is aligned to the *origin* of the **world coordinate system**.

4. Inside the *browser* window, click on the **light bulb** icon in front of the **origin** item to show the origin items in the graphics window.

5. Inside the *browser* window, click once with the right-mouse-button on the **XY Plane** item to display the option menu. Click on **Show/Hide** to toggle on the display of the datum features.

6. On your own, repeat the above steps and toggle ***ON/OFF*** the display of the individual *work planes*, *work axes*, and the *origin point* on the screen.

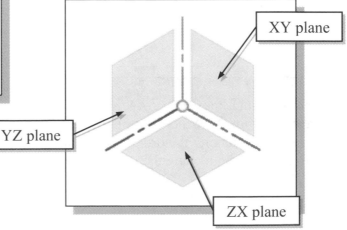

XY plane

YZ plane

ZX plane

❖ By default, the basic set of work planes is aligned to the world coordinate system; the work planes are the first features of the part. We can now proceed to create solid features referencing the three mutually perpendicular datum planes.

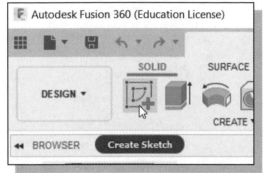

7. In the *Create panel* select the **Create Sketch** command by left-clicking once on the icon.

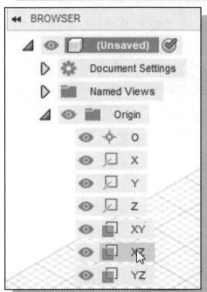

8. Move the cursor on top of the **XZ Plane**, inside the *browser* window as shown, and notice that Autodesk Fusion 360 will automatically highlight the corresponding plane in the graphics window. Left-click once to select the XZ Plane as the sketching plane.

❖ Autodesk Fusion 360 allows us to identify and select features in the graphics window as well as in the *browser* window.

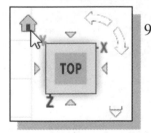

9. Single left-click to activate the **Home View** option as shown. The view will be adjusted back to the default *isometric view*.

➢ Note the alignment of the sketch plane is set to the XZ plane as shown.

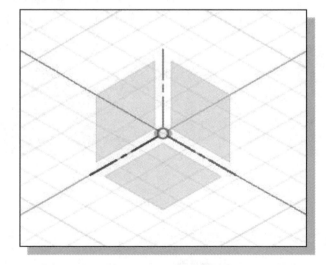

Create the 2D Sketch for the Base Feature

1. Select the **Center to center Slot** command by clicking once with the **left-mouse-button** on the icon in the *Create panel*.

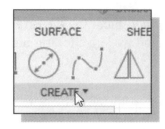
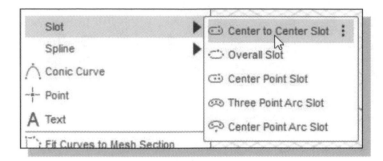

2. Create a horizontal line, representing center to center distance of the slot, in front of the origin as shown below. (Note the Horizontal constraint showing the alignment of the line.)

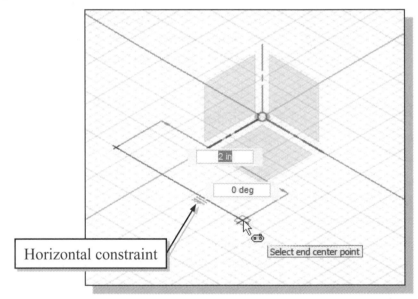

Horizontal constraint

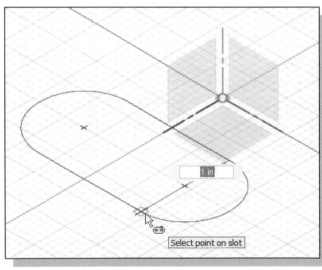

3. Move the cursor outward to define the size of the center to center slot; click once with the left-mouse-button to create the *Center to Center Slot*.

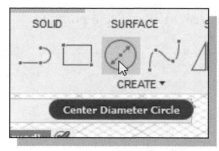

4. Choose **Center Diameter Circle** in the *Create panel*.

5. On your own, create the two inner circles; the center points of the circles are coincident to the centers of the oval slot.

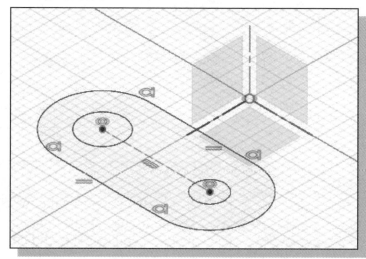

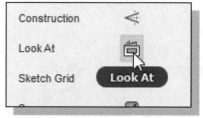

6. In the *Sketch Palette* panel, use the **Look At** command and set the display orientation to the **Top** view position as shown in the below figure.

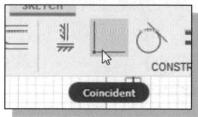

7. Click on the **Coincident** constraint icon in the *Sketch Palette* panel.

8. On your own, apply a **coincident constraint** between the origin and the center line as shown.

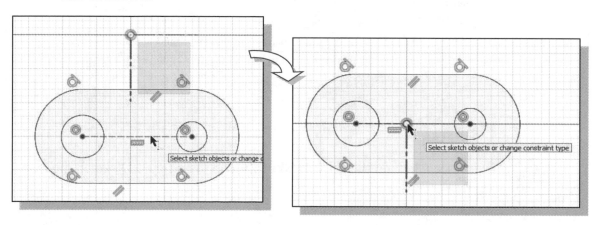

9. On your own, create and adjust the dimensions, referencing the origin point, to fully constrain the sketch as shown.

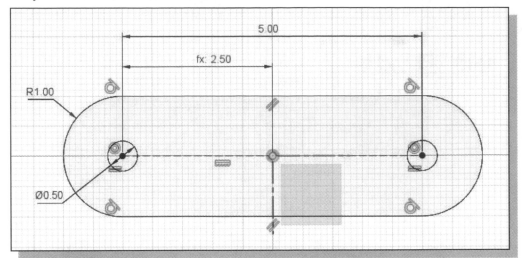

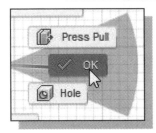

10. Inside the graphics window, click once with the right-mouse-button to display the option menu. Select **OK** in the pop-up menu to end the Sketch Dimension command.

11. In the *Modify* tool panel select the **Change Parameters** command by left-clicking once on the icon.

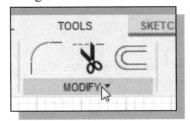
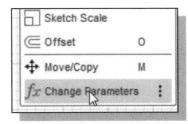

12. On your own, examine the parametric equation used.

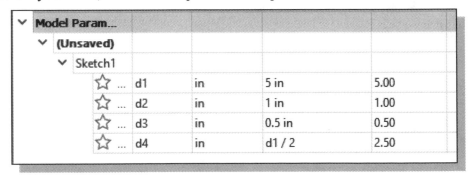

∨ **Model Param...**					
	∨ **(Unsaved)**				
		∨ Sketch1			
		☆ ...	d1	in	5 in	5.00
		☆ ...	d2	in	1 in	1.00
		☆ ...	d3	in	0.5 in	0.50
		☆ ...	d4	in	d1 / 2	2.50

13. Click on the **Finish Sketch** icon to exit the 2D sketch mode.

Create the First Extrude Feature

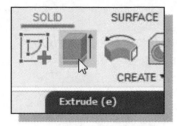

1. In the *Create panel*, select the **Extrude** command by left-clicking once on the icon.

2. Select the inside region of the sketch to define the profile of the extrusion as shown.

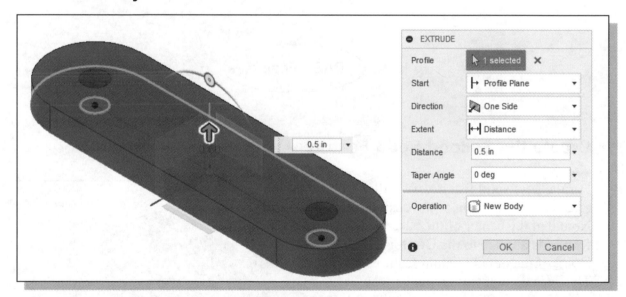

3. In the *Distance* option box, enter **0.5** as the extrusion distance and *operation* to **New Body** as shown.

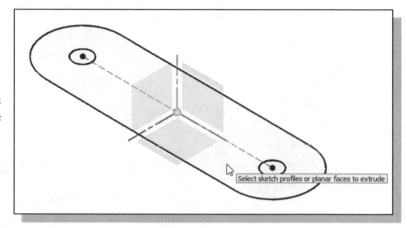

4. In the *Extrude* control panel, click on the **OK** button to create the base feature.

The Implied Parent/Child Relationships

- The *Model Tree* shows two main features: the ***Origin*** (default datum features) and the base feature (**Extrude1** shown in the figure) we just created. The parent/child relationships were established implicitly when we created the base feature: (1) XZ workplane was selected as the sketch plane; (2) Center Point was used as the reference point to align the 2D sketch of the base feature.

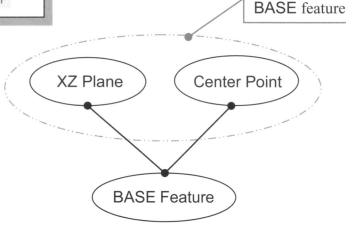

Create the Second Solid Feature

For the next solid feature, we will create the top section of the design. Note that the center of the base feature is aligned to the Center Point of the default datum features. This was done intentionally so that additional solid features can be created referencing the default datum features. For the second solid feature, the XY workplane will be used as the sketch plane.

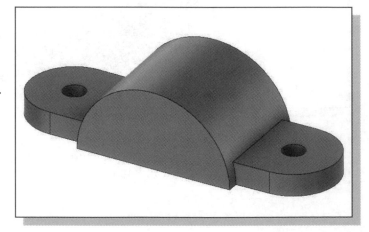

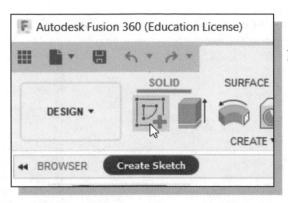

1. In the *Create panel* select the **Create Sketch** command by left-clicking once on the icon.

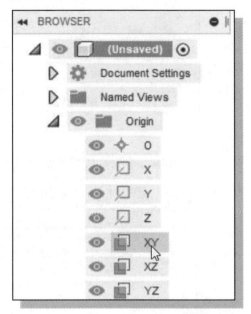

2. In the *Status Bar* area, the message "*Select face, work plane, sketch or sketch geometry.*" is displayed. Pick the **XY Plane** by clicking the work plane name inside the *browser* as shown.

3. Switch to one of the **Wireframe** display modes under *Visual Style* as shown.

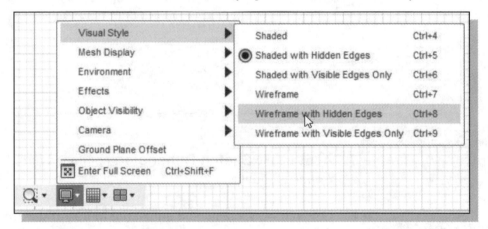

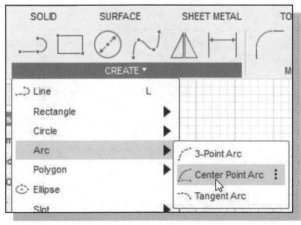

4. Select the **Center point arc** command by clicking once with the left-mouse-button on the icon in the *Create panel* as shown.

5. Pick the **origin** and watch for the alignment as the center location of the new arc.

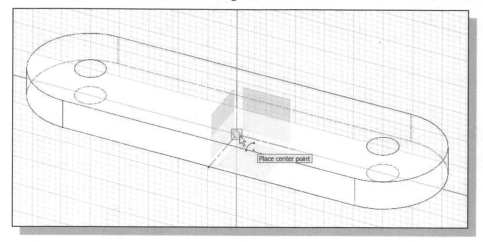

6. On your own, create a semi-circle of arbitrary size, with both endpoints aligned to the X-axis, as shown below.

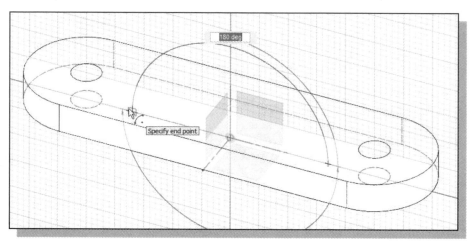

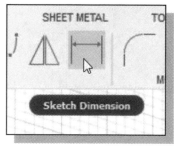

7. Activate the **Sketch Dimension** command in the *Create Panel* as shown.

8. On your own, create and adjust the radius of the arc to **1.75**.

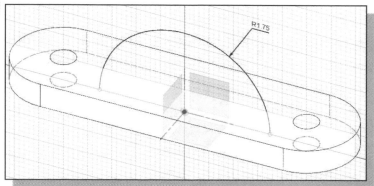

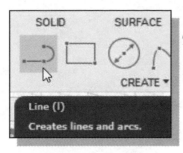

9. Select the **Line** command in the *Create panel*.

10. Create a line connecting the two endpoints of the arc as shown in the figure below. (Hint: apply a **Coincident** constraint to help align the line to the origin.)

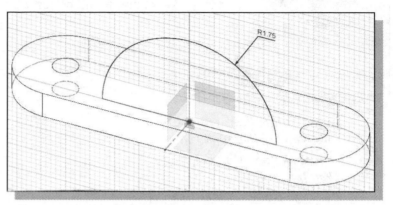

11. Select **Finish Sketch** in the *Ribbon* toolbar area to end the Sketch option.

12. In the *Create panel,* activate the **Extrude** command by left-clicking once on the icon.

13. Select the inside region of the sketched arc-line curves as the profile to be extruded.

14. In the *Extrude* dialog box, set to use the **Symmetric** and **Join** options. In the *Distance* value box, set the extrusion distance to **total width 2.5**.

15. Click on the **OK** button to create the feature.

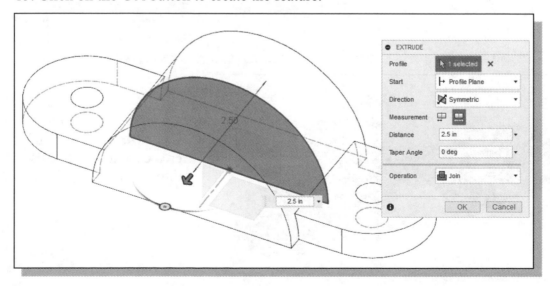

Create the Rectangular Cut Feature

- A rectangular cut will be created as the next solid feature.

1. In the *Create panel* select the **Create Sketch** command by left-clicking once on the icon.

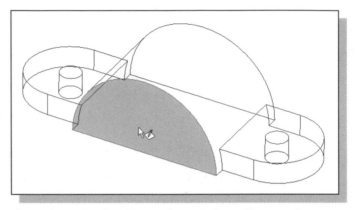

2. In the *Status Bar* area, the message "*Select plane to create sketch or an existing sketch to edit*" is displayed. Pick the front vertical face of the solid model shown.

3. On your own, create a rectangle and apply the dimensions (**size dimensions: 2.0x1.0**) as shown below. (Hint: Use the **origin** to align the location dimension.)

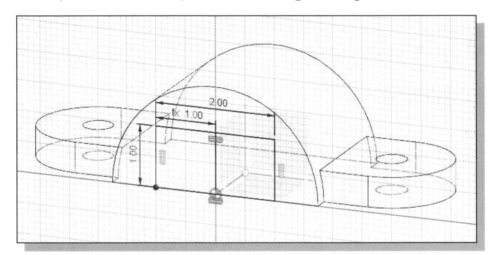

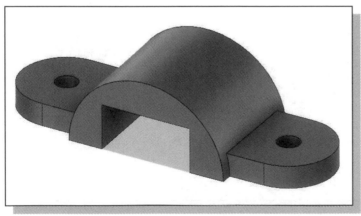

4. On your own, use the **Extrude** command and create a cutout that cuts through the entire 3D solid model as shown.

The Second Cut Feature

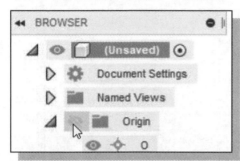

1. *Left-click* on the eye ball symbol to turn **OFF** the visibility of the datum features as shown.

2. In the *Create panel* select the **Create Sketch** command by left-clicking once on the icon.

3. Select the horizontal face of the last cut feature as the *sketching plane*.

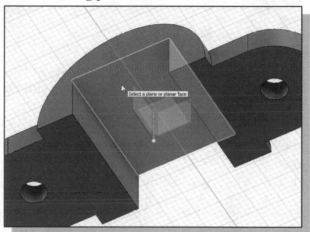

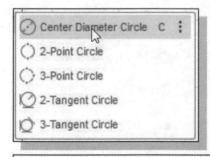

4. Select the **Center Diameter Circle** command by clicking once with the left-mouse-button on the icon in the *Create panel*.

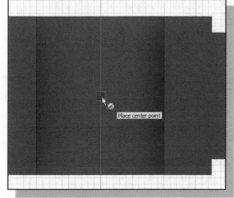

5. Pick the **origin**, located at the center of the part, to align the center of the new circle.

6. On your own, create a circle of arbitrary size.

7. On your own, add the size dimension of the circle and set the dimension to **0.25**.

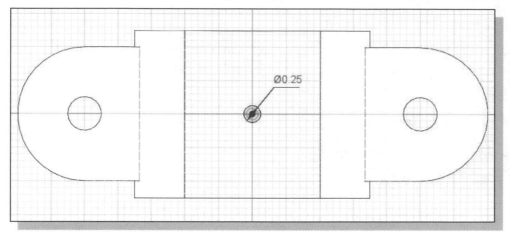

8. Click on the **Finish Sketch** icon to exit the 2D sketch mode.

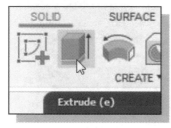

9. In the *Create toolbar*, select the **Extrude** command by clicking once with the left-mouse-button on the icon.

10. Select the inside region of the sketched circle as the profile to be extruded.

11. Drag the arrow control to set the extrusion direction upward. In the *Extrude* control panel, set to the **Cut - All** option.

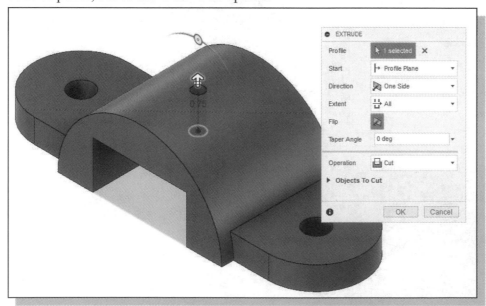

12. Click on the **OK** button to proceed with creating the cut feature.

Examine the Parent/Child Relationships

1. On your own, rename the feature names to **Base**, **MainBody**, **Rect_Cut** and **Center_Drill** as shown in the *Timeline control panel*.

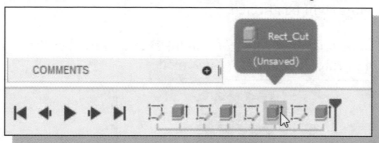

❖ The *Model Tree* window now contains seven items: the ***Origin*** (default datum features) and four solid features. All of the parent/child relationships were established implicitly as we created the solid features. As more features are created, it becomes much more difficult to make a sketch showing all the parent/child relationships involved in the model. On the other hand, it is not really necessary to have a detailed picture showing all the relationships among the features. In using a feature-based modeler, the main emphasis is to consider the interactions that exist between the **immediate features**. Treat each feature as a unit by itself, and be clear on the parent/child relationships for each feature. Thinking in terms of *features* is what distinguishes *feature-based modeling* and the previous generation solid modeling techniques. Let us take a look at the last feature we created, the **Center_Drill** feature. What are the parent/child relationships associated with this feature? (1) Since this is the last feature we created, it is not a parent feature to any other features. (2) Since we used one of the surfaces of the rectangular cutout as the sketching plane, the **Rect_Cut** feature is a parent feature to the **Center_Drill** feature. (3) We also used the Origin as a reference point to align the sketch; therefore, the ***Origin*** is also a parent to the **Center_Drill** feature.

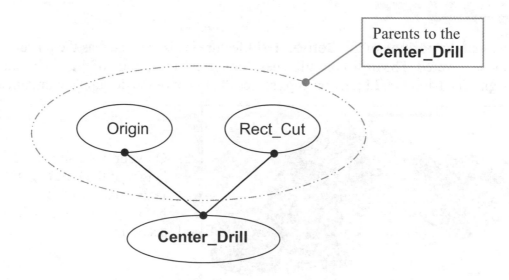

Parents to the **Center_Drill**

Modify a Parent Dimension

Any changes to the parent features will affect the child feature. For example, if we modify the height of the **Rect_Cut** feature from 1.0 to 0.75, the depth of the child feature (**Center_Drill** feature) will be affected.

1. In the *Timeline Control panel*, *double click* on the sketch in front of the **Rect_Cut** feature to enter the *Edit 2D sketch* mode.

2. Select the height dimension (1.0) by double-clicking on the dimension.

3. Enter **0.75** as the new height dimension as shown.

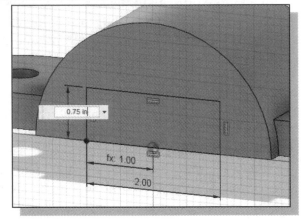

4. Click on the **Finish Sketch** button to proceed with updating the solid model.

➢ Note that the position of the **Center_Drill** feature is also adjusted as the placement plane is lowered. The drill-hole still goes through the main body of the *U-Bracket* design. The parent/child relationship assures the intent of the design is maintained.

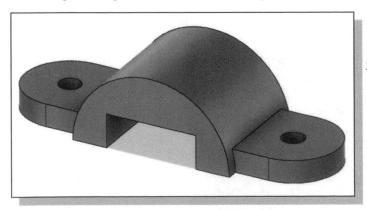

5. On your own, adjust the height of the **Rect_Cut** feature back to **1.0** inch before proceeding to the next section.

A Design Change

Engineering designs usually go through many revisions and changes. For example, a design change may call for a circular cutout instead of the current rectangular cutout feature in our model. Autodesk Fusion 360 provides an assortment of tools to handle design changes quickly and effectively. In the following sections, we will demonstrate some of the more advanced tools available in Autodesk Fusion 360, which allow us to perform the modification of changing the rectangular cutout (2.0 × 1.0 inch) to a circular cutout (radius: 1.25 inch).

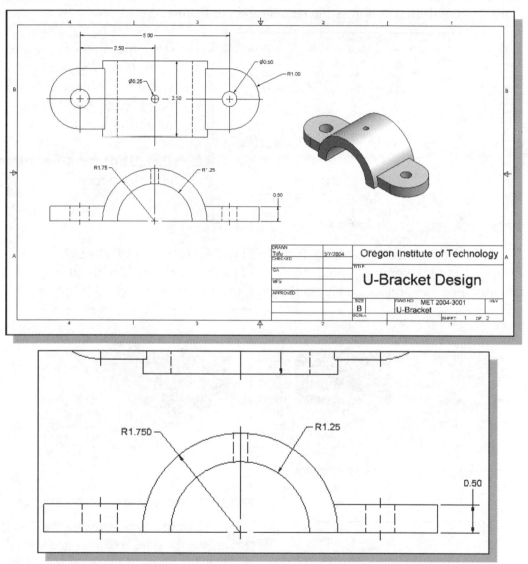

* ❖ Based on your knowledge of Autodesk Fusion 360 so far, how would you accomplish this modification? What other approaches can you think of that are also feasible? Of the approaches you came up with, which one is the easiest to do and which is the most flexible? If this design change was anticipated right at the beginning of the design process, what would be your choice in arranging the order of the features? You are encouraged to perform the modifications prior to following through the rest of the tutorial.

Feature Suppression

With Autodesk Fusion 360, we can take several different approaches to accomplish this modification. We could (1) create a new model, or (2) change the shape of the existing cut feature using the **Redefine** command, or (3) perform **feature suppression** on the rectangular cut feature and add a circular cut feature. The third approach offers the most flexibility and requires the least amount of editing to the existing geometry. **Feature suppression** is a method that enables us to disable a feature while retaining the complete feature information; the feature can be reactivated at any time. Prior to adding the new cut feature, we will first suppress the rectangular cut feature.

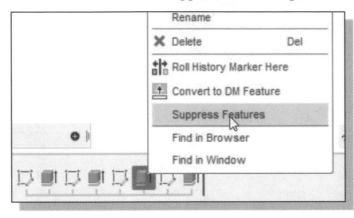

1. Move the cursor to the *Timeline control panel* area; click once with the right-mouse-button on top of **Rect_Cut** to bring up the option menu.

2. Click **Suppress Features** in the pop-up menu.

❖ With the *Suppress* command, the Rect_Cut and Center_Drill features have disappeared in the graphics area. The child feature cannot exist without its parent(s), and any modification to the parent (Rect_Cut) influences the child (Center_Drill).

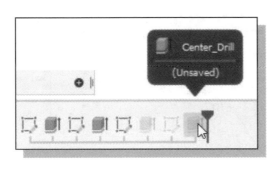

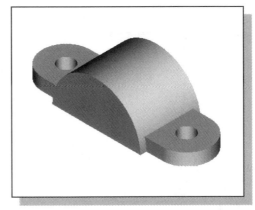

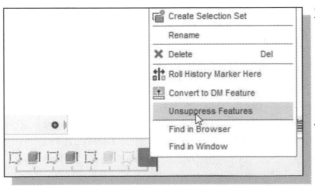

3. Move the cursor inside the *Timeline control panel*. Click once with the right-mouse-button on top of **Center_Drill** to bring up the option menu.

4. Pick **Unsuppress Features** in the pop-up menu to activate the suppressed features.

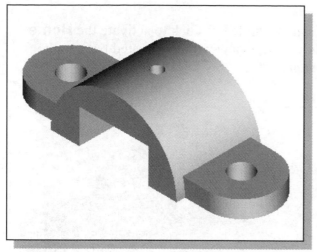

> ➢ In the graphics area, both the **Rect_Cut** feature and the **Center_Drill** feature are re-activated. The child feature cannot exist without its parent(s); the parent (Rect_Cut) must be activated to enable the child (Center_Drill).

A Different Approach to the Center_Drill Feature

❖ The main advantage of using the BORN technique is to provide greater flexibility for part modifications and design changes. In this case, the Center_Drill feature can be placed on the XZ workplane and therefore not be linked to the Rect_Cut feature.

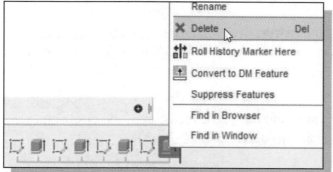

1. Move the cursor inside the *Timeline control panel*. Click once with the **right-mouse-button** on top of **Center_Drill** to bring up the option menu.

2. Pick **Delete** in the pop-up menu.

3. On your own, repeat the above steps and delete Sketch4.

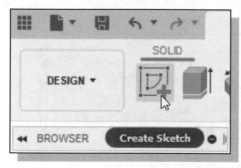

4. In the *Create panel* select the **Create Sketch** command by left-clicking once on the icon.

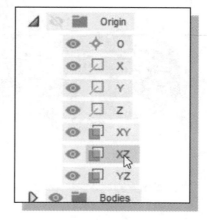

5. Pick the **XZ Workplane** in the *Browser* window as shown.

6. Inside the graphics area, single left-click to activate the **Home View** option as shown. The view will be adjusted back to the default *isometric view*.

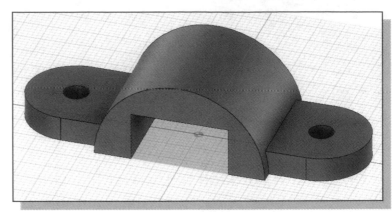

➢ Note the alignment of the sketch plane is set to the XZ plane as shown.

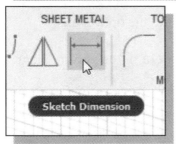

7. Select the **Center Diameter Circle** command by clicking once with the left-mouse-button on the icon in the *Create panel*.

8. Pick the **origin** to align the center of the new circle. Select another location to create a circle of arbitrary size.

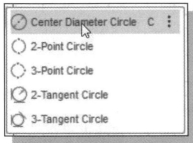

9. On your own, add the size dimension of the circle and set the dimension to **0.25**.

10. On your own, complete the extrude cut feature, cutting upward through the main body of the design.

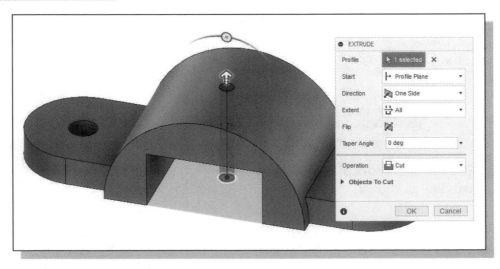

Suppress the Rect_Cut Feature

Now that the **Center_Drill** feature is no longer a child of the **Rect_Cut** feature, any changes to the **Rect_Cut** feature do not affect the **Center_Drill** feature.

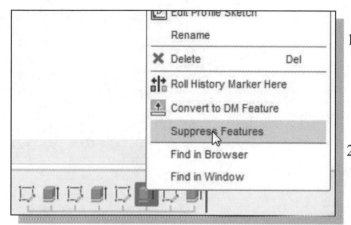

1. Move the cursor to the *Timeline control panel* area; click once with the right-mouse-button on top of **Rect_Cut** to bring up the option menu.

2. Pick **Suppress Features** in the pop-up menu.

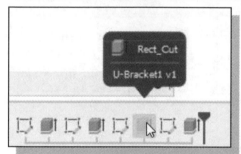

❖ The **Rect_Cut** feature is now disabled without affecting the *new* **Center_Drill** feature.

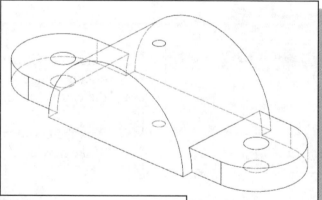

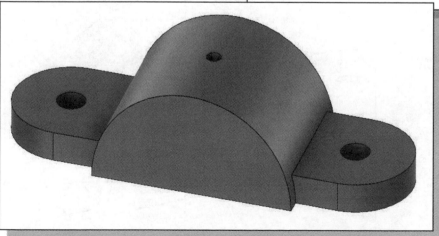

Create a Circular Cut Feature

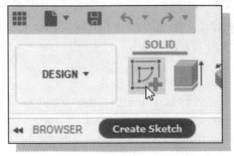

1. In the *Create panel,* select the **Create Sketch** command by left-clicking once on the icon.

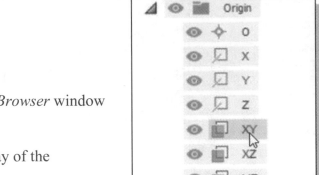

2. Pick the XY workplane in the *Browser* window as shown.

3. On your own, turn on the display of the coordinate system.

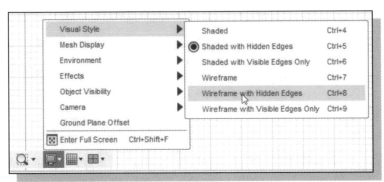

4. Switch to one of the **Wireframe** display modes under *Visual Style* as shown.

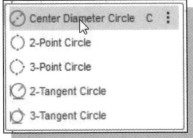

5. Select the **Center Diameter Circle** command by clicking once with the left-mouse-button on the icon in the *Create panel.*

6. Pick the **center point** at the origin to align the center of the new circle.

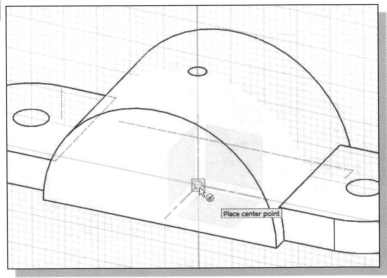

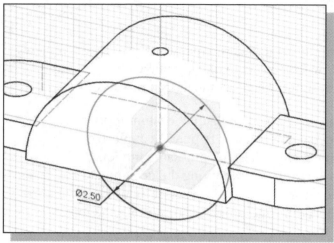

7. On your own, create a circle of diameter **2.5** as shown in the figure.

8. On your own, complete the **cut** feature using the **Symmetric** and **distance** of **2.5**.

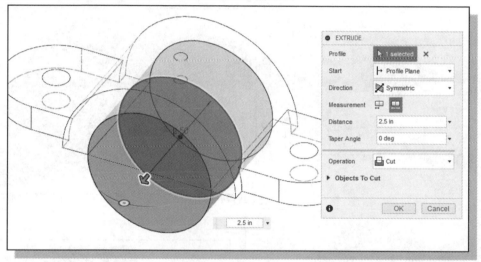

❖ Note that the parents of the Circular_Cut feature are the XY Plane and the Center Point.

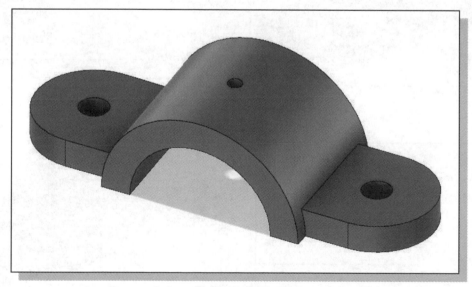

➢ On your own, save the model as ***U-Bracket***; this model will be used again in the next chapter.

A Flexible Design Approach

In a typical design process, the initial design will undergo many analyses, tests, reviews and revisions. Autodesk Fusion 360 allows the users to quickly make changes and explore different options of the initial design throughout the design process.

The model we constructed in this chapter contains two distinct design options. The *feature-based parametric modeling* approach enables us to quickly explore design alternatives and we can include different design ideas into the same model. With parametric modeling, designers can concentrate on improving the design, and the design process becomes quicker and more effortless. The key to successfully using parametric modeling as a design tool lies in understanding and properly controlling the interactions of features, especially the parent/child relations.

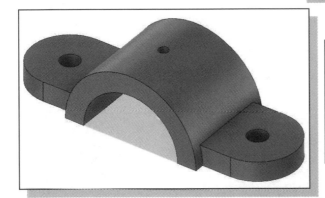
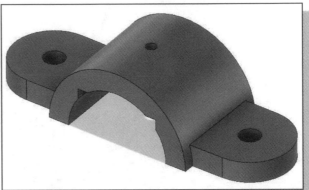
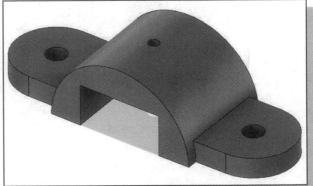
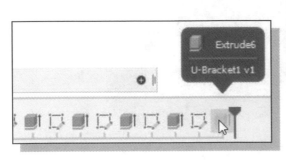
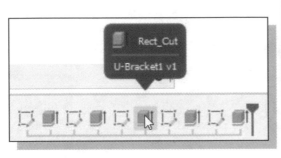
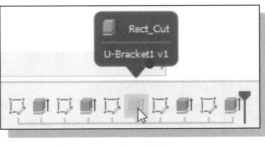

View and Edit Material Properties

The *Fusion 360 Material Library* provides many commonly used materials. The material properties listed in the *Material Library* can be edited and new materials can be added.

1. In the **Modify** panel, click **Manage Materials** as shown.

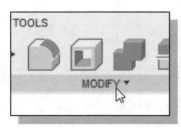 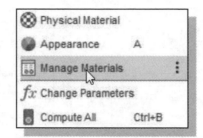

2. In the *Fusion 360 Material* library group, select **Aluminum-6061**.

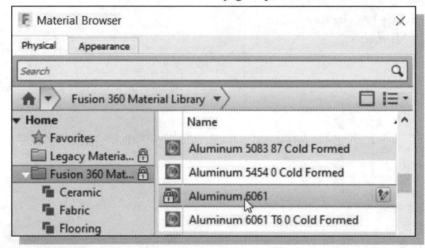

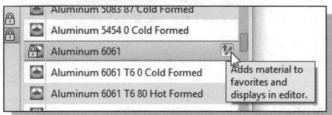

3. Click the **Add Material to favorites and display in editor** icon as shown.

4. Click on the **Physical Aspect** tab to view the associated material properties.

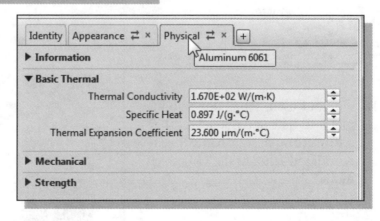

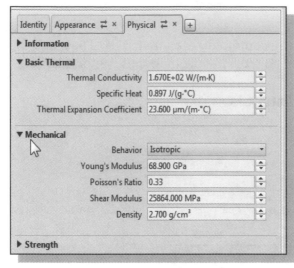

5. Expand the **Mechanical properties** list as shown.

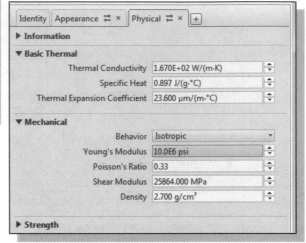

6. Enter **10.0E6 psi** as the new value for *Young's Modulus*.

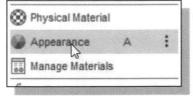

7. Click **OK** to accept the changes and close the editor.

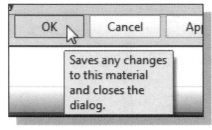

8. Click **Close** to exit the **Material Browser**.

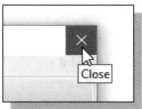

9. In the **Modify** panel, click **Appearance** as shown.

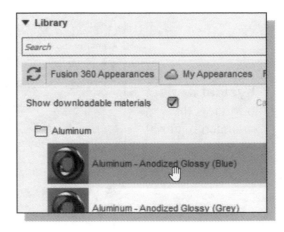

10. On your own, scroll down in the material lists and open **Metal →Aluminum**; assign **Aluminum-Anodized Blue** to the U-bracket model.

Predefined keyboard and mouse shortcuts in Fusion360:

Function	Windows	Mac
Model workspace		
Extrude	E	E
Hole	H	H
Press/Pull	Q	Q
Fillet	F	F
Move	M	M
Appearance	A	A
Compute All	Ctrl+B	Command+B
Joint	J	J
As-built Joint	Shift+J	Shift+J
Line	L	L
2-point Rectangle	R	R
Center Diameter Circle	C	C
Trim	T	T
Offset	O	O
Measure	I	I
Sketch Dimension	D	D
Scripts and Add-Ins	Shift+S	Shift+S
Window Selection	1	1
Freeform Selection	2	2
Paint Selection	3	3
Mouse/Touchpad:		
Pan	press middle button	press middle button two-finger drag
Zoom	roll wheel	roll wheel pinch
Orbit	Shift + press middle button	Shift + press middle button Shift + two-finger drag
Orbit around point	Shift + click then press middle button	Shift + click then press middle button

Review Questions:

1. Why is it important to consider the parent/child relationships between features?

2. Describe the procedure to suppress a feature.

3. What is the basic concept of the BORN technique?

4. What happens to a feature when it is suppressed?

5. How do you identify a suppressed feature in a model?

6. What is the main advantage of using the BORN technique?

7. Create sketches showing the steps you plan to use to create the models shown on the next page:

Exercises: Create and save the exercises in the Chapter7 folder.

1. **Swivel Yoke** (Dimensions are in inches, Material: **Cast Iron**)

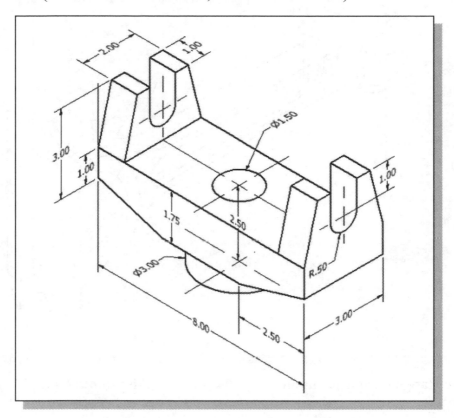

2. **Angle Bracket** (Dimensions are in inches, Material: **Carbon Steel**)

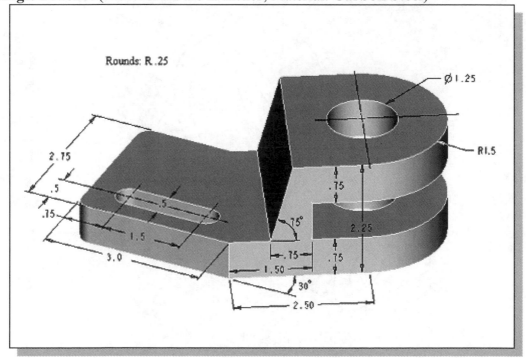

3. **Connecting Rod** (Dimensions are in inches, Material: **Carbon Steel**)

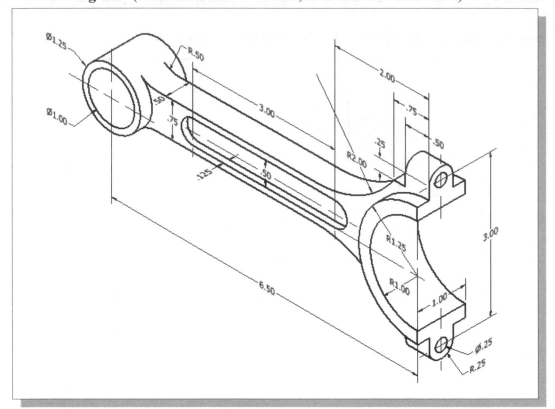

4. **Tube Hanger** (Dimensions are in inches, Material: **Aluminum 6061**)

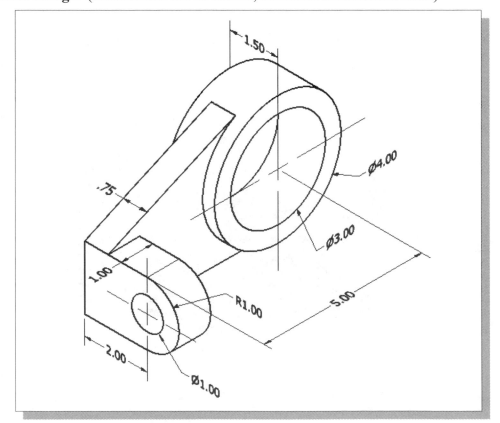

5. **Angle Latch** (Dimensions are in millimeters, Material: **Brass**)

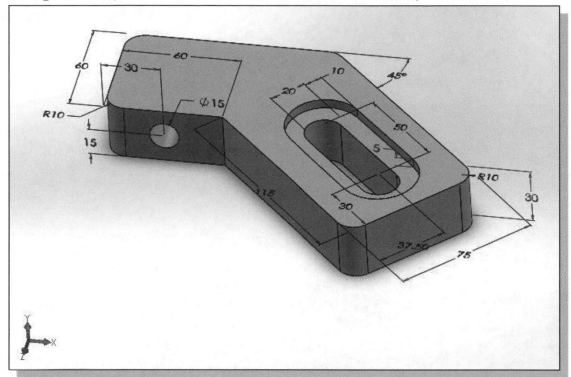

6. **Inclined Lift** (Dimensions are in inches, Material: **Mild Steel**)

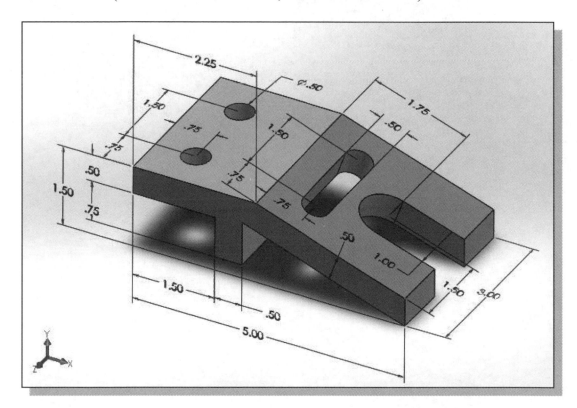

Notes:

Chapter 8
Part Drawings and Associative Functionality

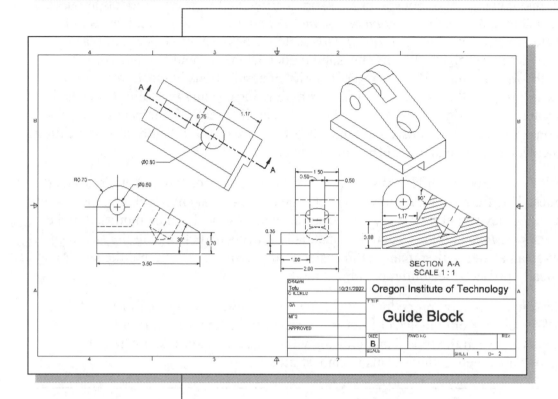

SECTION A-A
SCALE 1 : 1

DRAWN			
Tofu C. BLOKLU	10/31/2002	Oregon Institute of Technology	
QA		TITLE	
MFG		**Guide Block**	
APPROVED			
SIZE		DWG NO	REV
B			
SCALE		SHEET 1 OF 2	

Learning Objectives

- ♦ **Create Drawing Layouts from Solid Models**
- ♦ **Understand Associative Functionality**
- ♦ **Use the Default Borders and Title Block in the Layout Mode**
- ♦ **Arrange and Manage 2D Views in Drawing Mode**
- ♦ **Display and Hide Feature Dimensions**
- ♦ **Create Reference Dimensions**
- ♦ **Create 3D Annotations in Isometric Views**

Drawings from Parts and Associative Functionality

With the software/hardware improvements in solid modeling, the importance of two-dimensional drawings is decreasing. Drafting is considered one of the downstream applications of using solid models. In many production facilities, solid models are used to generate machine tool paths for *computer numerical control* (CNC) machines. Solid models are also used in *rapid prototyping* to create 3D physical models out of plastic resins, powdered metal, etc. Ideally, the solid model database should be used directly to generate the final product. However, the majority of applications in most production facilities still require the use of two-dimensional drawings. Using the solid model as the starting point for a design, solid modeling tools can easily create all the necessary two-dimensional views. In this sense, solid modeling tools are making the process of creating two-dimensional drawings more efficient and effective.

Autodesk Fusion 360 provides associative functionality in the different Autodesk Fusion 360 modes. This functionality allows us to change the design at any level, and the system reflects it at all levels automatically. For example, a solid model can be modified in the *Part Modeling Mode* and the system automatically reflects that change in the *Drawing Mode*. We can also modify a feature dimension in the *Drawing Mode*, and the system automatically updates the solid model in all modes.

In this lesson, the general procedure of creating multi-view drawings is illustrated. The *U_Bracket* design from the last chapter is used to demonstrate the associative functionality between the model and drawing views. The procedure to create 3D annotations on isometric views is also demonstrated.

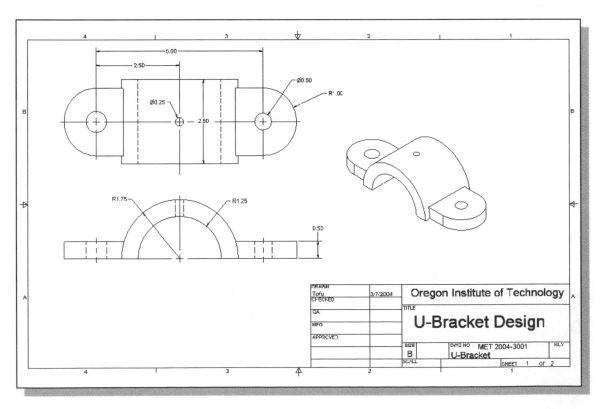

Starting Autodesk Fusion 360

1. Select the **Autodesk Fusion 360** option on the *Start* menu or select the **Autodesk Fusion 360** icon on the desktop to start Autodesk Fusion 360.

2. Click on the **Show Data Panel** icon as shown.

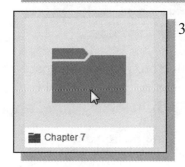

3. Open the *Parametric Modeling* project and double click on the **Chapter 7** folder.

4. Double click on the **U-Bracket** model to open the model as shown.

5. On your own, close the **Show Data Panel.**

Drawing Mode – 2D Paper Space

Autodesk Fusion 360 allows us to generate 2D engineering drawings from solid models so that we can plot the drawings to any exact scale on paper. An engineering drawing is a tool that can be used to communicate engineering ideas/designs to manufacturing, purchasing, service, and other departments. Until now we have been working in *model space* to create our design in *full size*. We can arrange our design on a two-dimensional sheet of paper so that the plotted hardcopy is exactly what we want. This two-dimensional sheet of paper is known as *paper space* in AutoCAD and Autodesk Fusion 360. We can place borders and title blocks, objects that are less critical to our design, on *paper space*. In general, each company uses a set of standards for drawing content, based on the type of product and also on established internal processes. The appearance of an engineering drawing varies depending on when, where, and for what purpose it is produced. However, the general procedure for creating an engineering drawing from a solid model is fairly well defined. In Autodesk Fusion 360, creation of 2D engineering drawings from solid models consists of four basic steps: drawing sheet formatting, creating/positioning views, annotations, and printing/plotting.

1. In the *File* pull down menu, select **New Drawing → From Design** to start a new drawing.

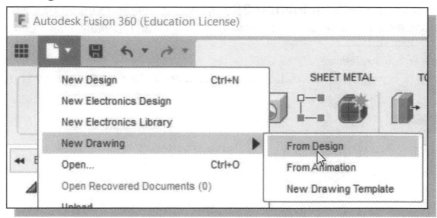

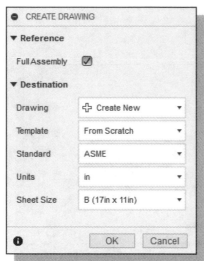

2. In the **Create Drawing** dialog box, set the Standard to ASME, units to in and sheet size to B (17in x 11 in) as shown in the figure. Click **OK** to proceed.

➢ Note that a new tab appears in the Fusion 360 window. We can switch between the solid model and the drawing by clicking the corresponding tabs in the ribbon toolbar area.

❖ In the graphics window, Autodesk Fusion 360 displays a default drawing sheet that includes a title block.

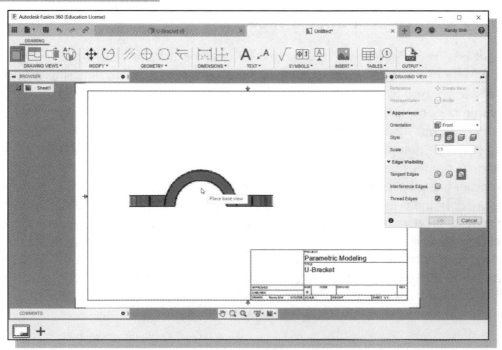

Add a Base View

In *Autodesk Fusion 360 Drawing Mode*, the first drawing view we create is called a **base view**. A *base view* is the primary view in the drawing; other views can be derived from this view. When creating a *base view*, Autodesk Fusion 360 allows us to specify the view to be shown. By default, Autodesk Fusion 360 will treat the *world XY plane* as the front view of the solid model. Note that there can be more than one *base view* in a drawing.

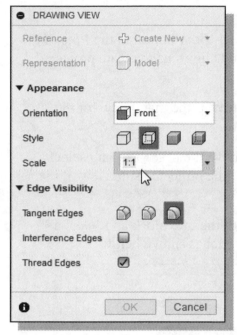

1. In the *Drawing View* dialog box, set the **Scale** to **1 : 1** and Style to **Hidden Line** as shown in the figure.

2. Confirm the *Tangent Edges* option is set to **off** as shown.

3. Inside the *graphics window,* drag and place the **base** view near the lower left corner of the graphics window as shown below. Click **OK** to place the *base view* and close the *Drawing View* dialog box.

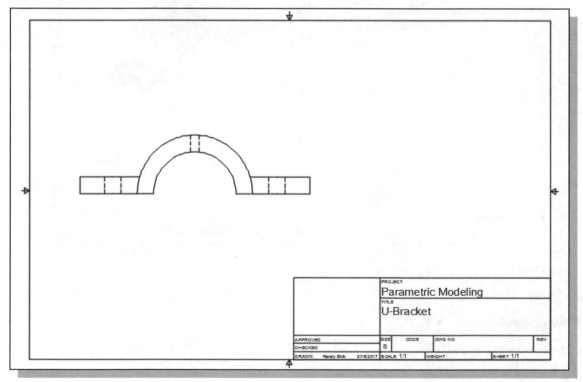

Create Projected Views

In *Autodesk Fusion 360 Drawing Mode*, **projected views** can be created with a first-angle or third-angle projection, depending on the drafting standard used for the drawing. We must have a base view before a projected view can be created. Projected views can be orthographic projections or isometric projections. Orthographic projections are aligned to the base view and inherit the base view's scale and display settings. Isometric projections are not aligned to the base view.

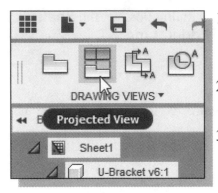

1. Click the **Projected View** button in the *Place Views* panel; this command allows us to create projected views.

2. Select the *base view* as the main view for the projected views.

3. Move the cursor above the base view and select a location to position the projected side view of the model.

4. Move the cursor toward the upper right corner of the title block and select a location to position the isometric view of the model as shown below.

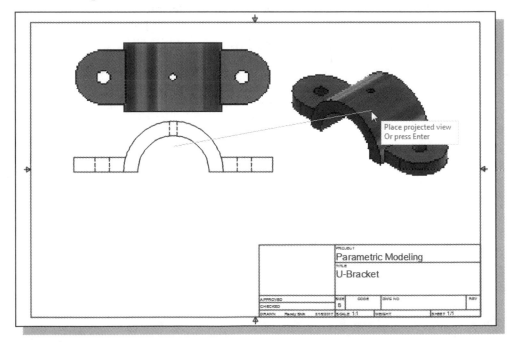

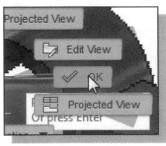

5. Inside the *graphics window*, right-click once to bring up the option menu.

6. Select **OK** to proceed with creating the two projected views and exit the *projected view* command.

Adjust the View Scale

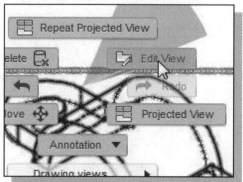

1. Select the isometric view; note the box around the entire view indicating the view is selected.

2. **Right-click** once to bring up the *option menu*.

3. Select **Edit View** in the option menu.

4. Select the *isometric view* to Edit.

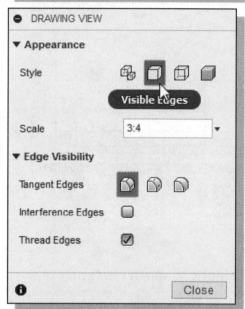

5. In the *Drawing View* dialog box, set the *Style* to **Visible edges** as shown.

6. Change the *Scale* to **3:4** as shown in the figure.

7. Set the *Tangent Edges* option to **off** as shown.

8. Click on the **Close** button to accept the settings and proceed with updating the drawing views.

Repositioning Views

1. Select the **front view** and notice the **Grip point** at the center.

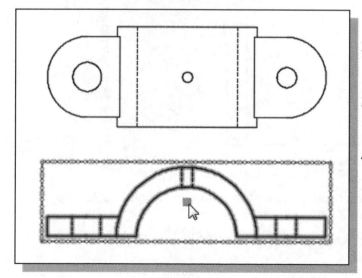

2. Click on the **grip point** to activate the Move option.

3. Move the cursor around to reposition the view.

4. Click the left-mouse-button again to reposition the view to a new location.

5. On your own, reposition the views we have created so far. Note that the top view can be repositioned only in the vertical direction. The top view remains aligned to the base view, the front view.

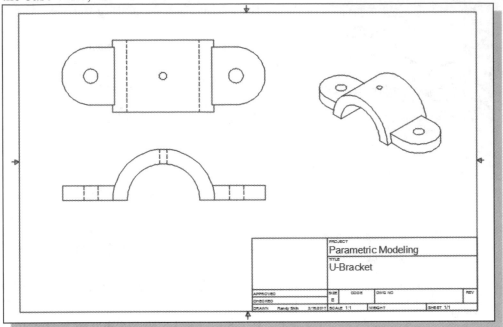

Add and adjust Center Marks

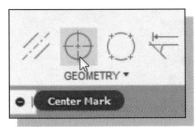

1. Click on the **Center Mark** button in the *Centerlines* panel as shown.

2. Click on the circle on the right in the top view to add a center mark as shown.

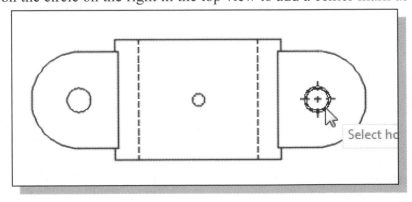

3. On your own, click on the other two circles in the top view to add center marks.

4. On your own, add a center mark in the front view by clicking on the large arc.

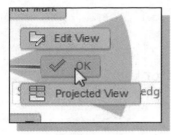

5. Inside the *graphics window*, right-click once to bring up the option menu.

6. Select **OK** to exit the *Center Mark* command.

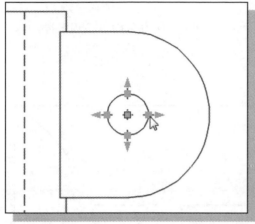

7. Click on the right centerlines in the *top* view as shown.

8. Adjust the length of the horizontal centerline by selecting one of the grey grip points as shown.

9. Click at a new location outward to lengthen the selected center line.

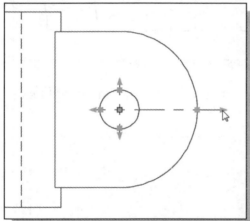

10. Click on one of the grey arrows to select it.

11. Move the cursor outward and select a new location. The grey arrow can also be used to adjust the appearance of the centerline.

12. On your own, repeat the above steps and adjust the centerlines in the top view as shown.

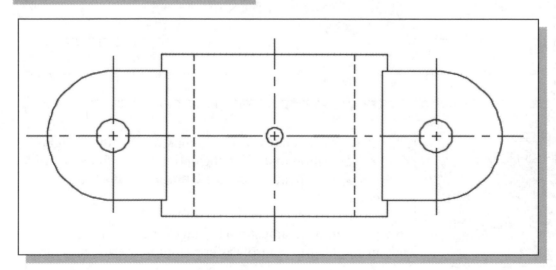

Add and adjust Centerline Bisectors

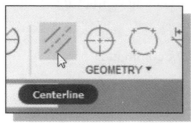

1. Click on the **Centerline Bisector** button in the *Centerlines* panel as shown.

2. Click on the two hidden edges of one of the *drill* features in the front view as shown in the figure.

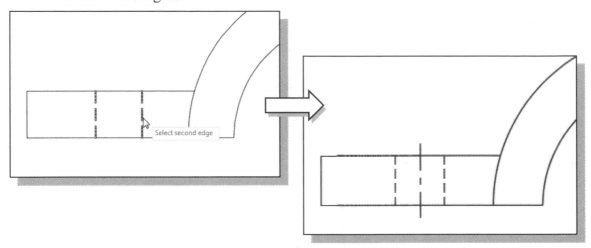

3. On your own, repeat the above step and create another centerline on the right side of the front view as shown.

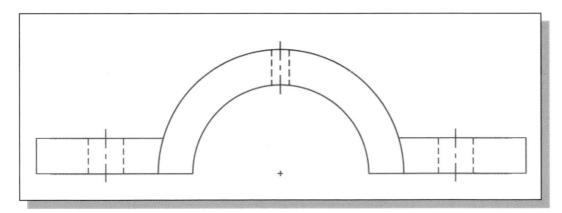

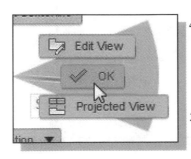

4. Inside the graphics window, click once with the right-mouse-button to display the option menu. Select **OK** in the pop-up menu to end the Centerline Bisector command.

5. On your own, adjust the length of the middle centerline using the grip controls as appeared in the above figure.

Dimensioning the Drawing

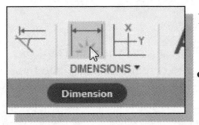

1. Select **Dimension** by left-clicking once in the *Dimensions* toolbar panel.

- The **Dimension** command in the drafting module is the same as the dimension command in the modeling module.

2. Select the center points of the circle on the left and the center point of the middle circle and place the dimension below the top view as shown.

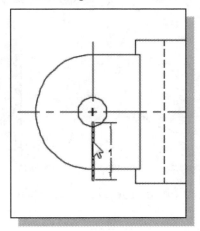

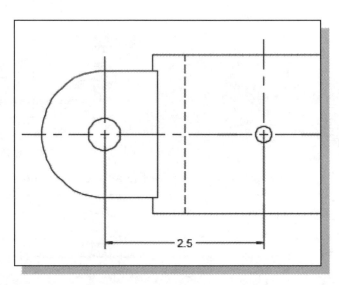

3. Repeat the above steps and create the center to center dimension between the left and right circles.

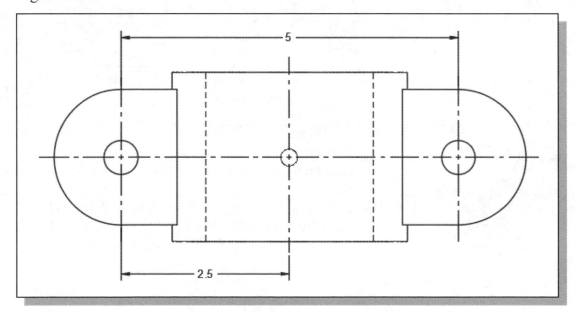

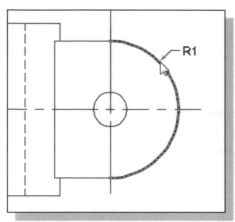

4. Select the arc on the right side to attach the leader to the circle as shown.

5. Move the cursor upward and toward the right, roughly at the 45-degree direction, and place the leader dimension as shown.

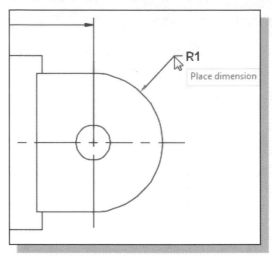

6. On your own, repeat the above steps and dimension the top view as shown.

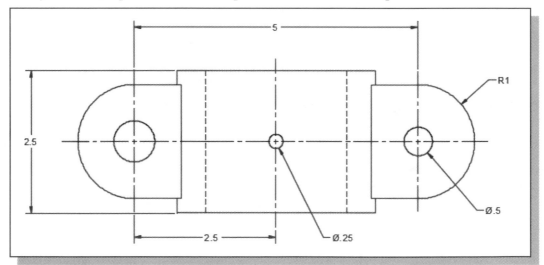

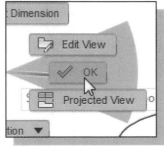

7. On your own, add dimensions to the front view.

8. Inside the graphics window, click once with the right-mouse-button to display the option menu. Select **OK** in the pop-up menu to end the Dimension command.

Repositioning Dimensions and 2D views

1. Select the width dimension **5.00**, and notice the different grip points available.

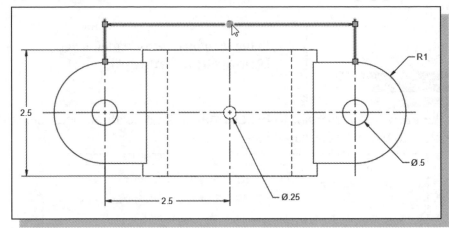

2. On your own, reposition the dimension text, using the left-mouse-button, to a new location.

3. Select the front view and notice the blue color inside dash box indicating the view is selected.

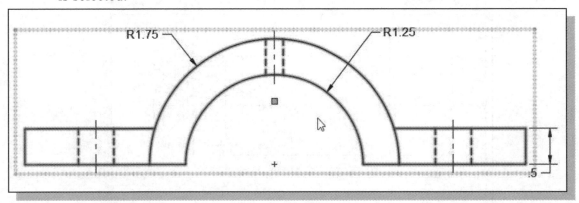

4. Using the grip point, reposition the view downward and note the location of all the associated dimensions is maintained.

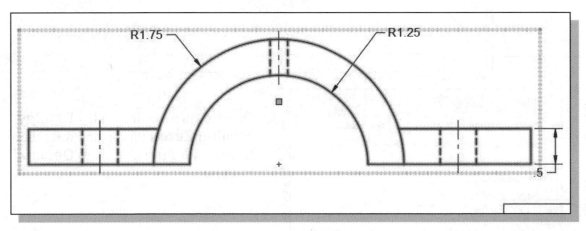

Drawing Dimensions Format

Autodesk Fusion 360 also provides controls to the drawing dimensions, such as font, precision, unit format.

1. In the *display control* panel, select **[Annotation Settings]**.

2. Set the *Linear Unit Format* to **Fractional** and notice all of the dimensions are switched to using fractional values.

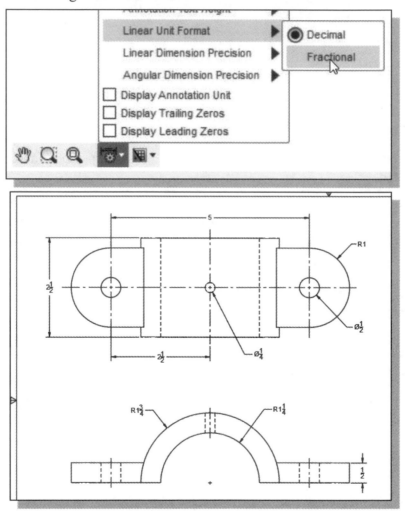

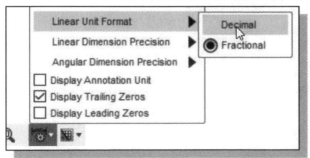

3. On your own, switch on the **Display Trailing Zeros** option and reset the *Linear Unit Format* back to **Decimal** as shown.

Complete the Drawing Sheet

1. On your own, use the **Zoom** and the **Pan** commands to adjust the display as shown; this is so we can complete the title block.

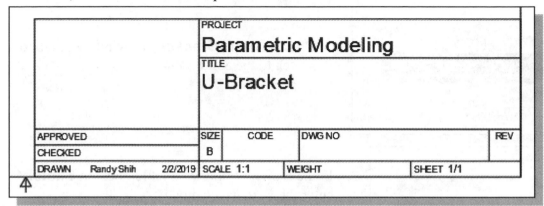

- Note that quite a few blocks are automatically filled by Autodesk Fusion 360, such as *Project Name*, *Part Title*, *Drawn By* information and *Scale*.

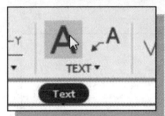

2. In the *Text Panel*, click on the **Text** button.

3. Pick a location that is inside the *DWG No* area and specify a text box region as shown.

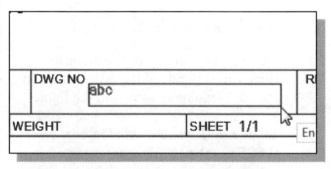

4. Enter **MECH 375-CH7-001** as the location for the new text to be entered.

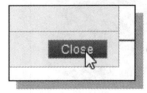

5. Click **Close** to accept the settings and end the *Text* command.

6. On your own, **Save** the drawing to the chapter 7 folder.

Associative Functionality – Modifying Feature Dimensions

Autodesk Fusion 360's *associative functionality* allows us to change the design and the system will reflect the changes at all levels.

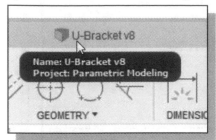

1. Click on the **U-Bracket** part window or tab to switch to the *Solid Model*.

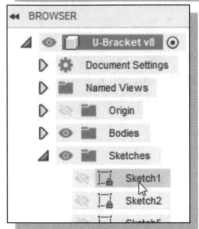

2. In the *browser* window, double-click on **Sketch1** to enter the *Edit Sketch* mode.

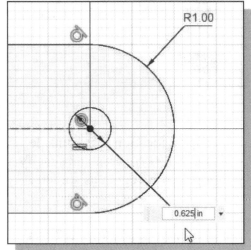

3. Double-click on one of the diameter dimensions (**0.50**) of the drill feature on the base feature as shown in the figure.

4. In the *Edit Dimension* dialog box, enter **0.625** as the new diameter dimension.

5. Click on the **Stop Sketch** button to update the model.

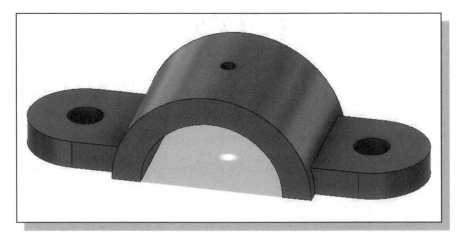

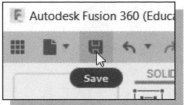

6. Click on the **Save** icon to accept the modification and upload the file to the cloud server.

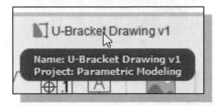

7. Click on the **U-Bracket** drawing graphics window or tab to switch to the *Multi-View Drawing*.

- Near the lower right corner of the screen, a warning message is present indicating the design has been modified and the drawing will need to be updated.

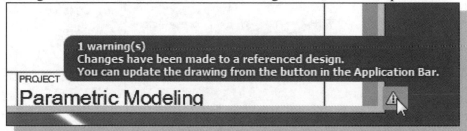

8. In the *Application bar*, click on the **Warning button** to proceed with updating the associated 2D drawing.

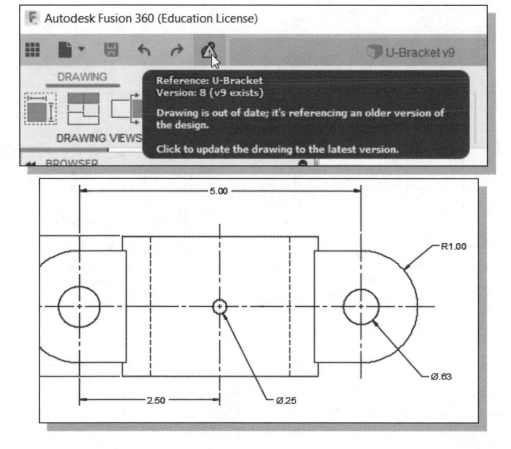

9. Double click on the **0.63** dimension to enter the *Edit Dimension* mode.

10. In the dimension dialog box, set the *Precision* option to **3 digits after the decimal point** as shown.

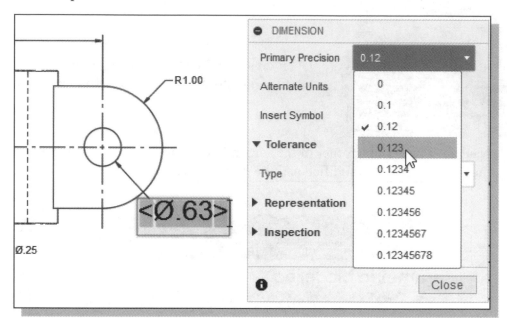

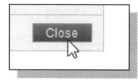

11. Click **Close** to accept the settings and end the *Edit dimension* option.

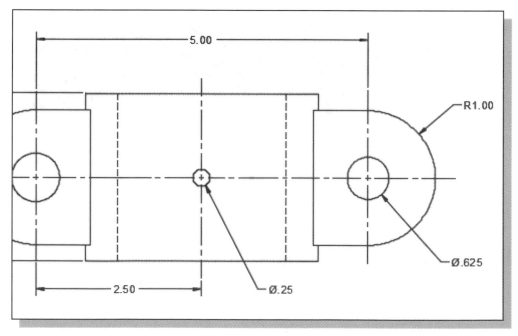

❖ On your own, complete the multi-view drawing, which should appear similar to the drawing shown on the next page.

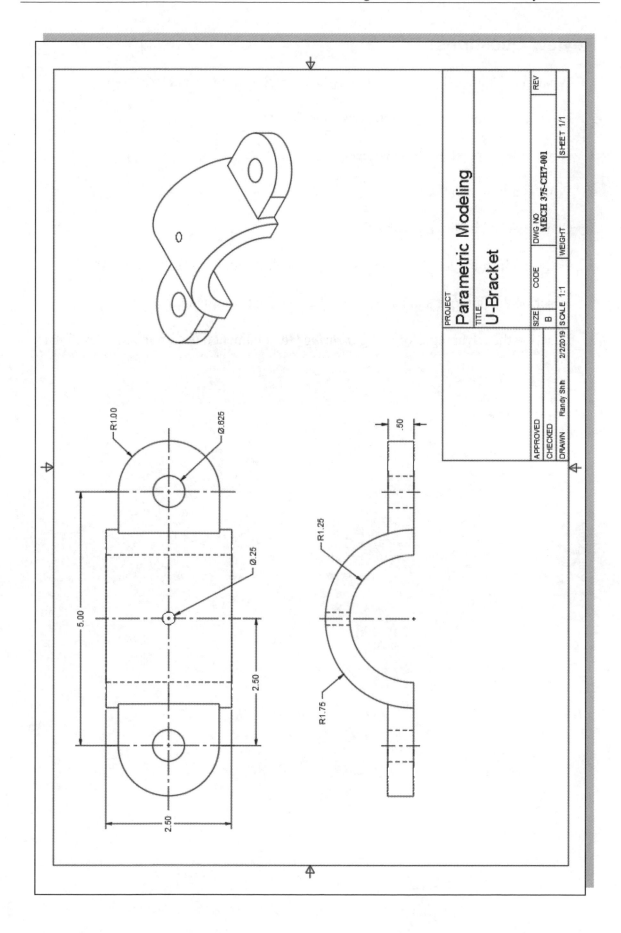

PROJECT	Parametric Modeling			REV
TITLE	U-Bracket			
SIZE B	CODE	DWG NO MECH 375-CH7-001		
SCALE 1:1		WEIGHT	SHEET 1/1	

APPROVED		
CHECKED		
DRAWN	Randy Shih	2/2/2019

R1.00
Ø.625
Ø.25
5.00
2.50
2.50
R1.25
R1.75
.50

Review Questions:

1. What does Autodesk Fusion 360's *associative functionality* allow us to do?

2. How do we move a view on the *Drawing Sheet*?

3. How do we reposition dimensions?

4. What is a *base view*?

5. Can we delete a drawing view? How?

6. Can we adjust the length of centerlines in the drafting mode of Fusion 360? How?

7. Describe the purpose and usage of the Leader Text command.

8. Describe the advantages of using *drawings* to document a design in Autodesk Fusion 360.

Exercises: Create the Solid models and the associated 2D drawings and save the exercises in the Chapter 8 folder.

1. **Slide Mount** (Dimensions are in inches.)

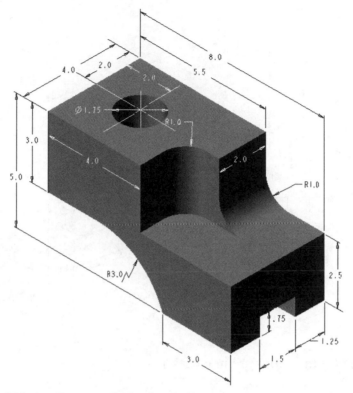

2. **Corner Stop** (Dimensions are in inches.)

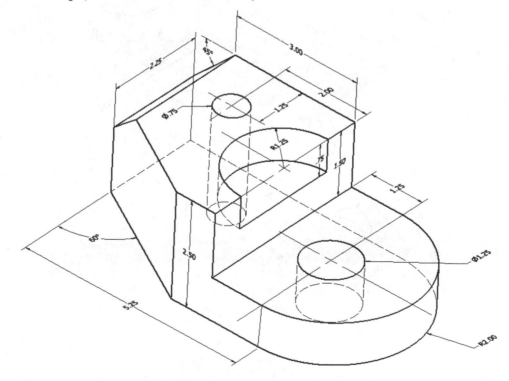

3. **Switch Base** (Dimensions are in inches.)

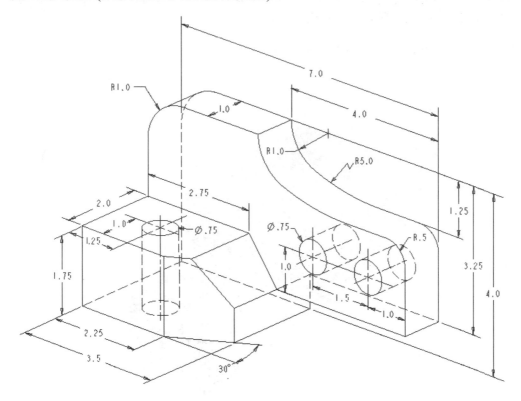

4. **Angle Support** (Dimensions are in inches.)

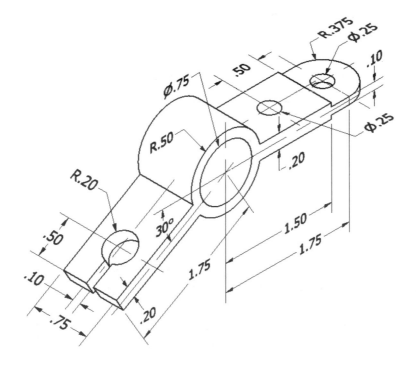

5. **Block Base** (Dimensions are in inches. Plate Thickness: 0.25)

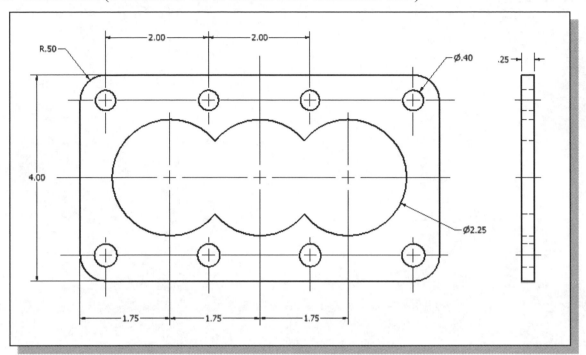

6. **Shaft Guide** (Dimensions are in inches.)

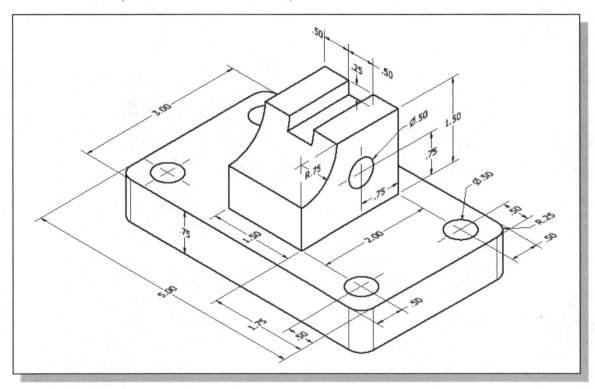

Notes:

Chapter 9
Datum Features and Auxiliary Views

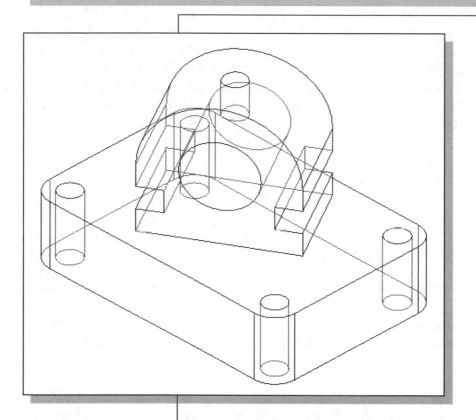

Learning Objectives

- **Understand the Concepts and the Use of Work Features**
- **Use the Different Options to Create Work Features**
- **Create Auxiliary Views in 2D Drawing Mode**
- **Create and Adjust Centerlines**
- **Create Shaded Images in the 2D Drawing Mode**

Work Features

Feature-based parametric modeling is a cumulative process. The relationships that we define between features determine how a feature reacts when other features are changed. Because of this interaction, certain features must, by necessity, precede others. A new feature can use previously defined features to define information such as size, shape, location and orientation. *Autodesk Fusion 360* provides several tools to automate this process. Work features can be thought of as user-definable datum, which are updated with the part geometry. We can create work planes, axes, or points that do not already exist. Work features can also be used to align features or to orient parts in an assembly. In this chapter, the use of the **Offset** option and the **Angled** option to create new work planes, surfaces that do not already exist, is illustrated. By creating parametric work features, the established feature interactions in the CAD database assure the capturing of the design intent. The default work features, which are aligned to the origin of the coordinate system, can be used to assist the construction of the more complex geometric features.

Auxiliary Views in 2D Drawings

An important rule concerning multiview drawings is to draw enough views to accurately describe the design. This usually requires two or three of the regular views, such as a front view, a top view and/or a side view. However, many designs have features located on inclined surfaces that are not parallel to the regular planes of projection. To truly describe the feature, the true shape of the feature must be shown using an **auxiliary view**. An *auxiliary view* has a line of sight that is perpendicular to the inclined surface, as viewed looking directly at the inclined surface. An *auxiliary view* is a supplementary view that can be constructed from any of the regular views. Using the solid model as the starting point for a design, auxiliary views can be easily created in 2D drawings. In this chapter, the general procedure of creating auxiliary views in 2D drawings from solid models is illustrated.

The Rod-Guide Design

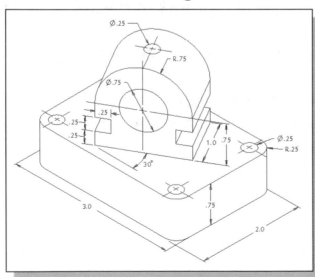

❖ Based on your knowledge of Autodesk Fusion 360 so far, how would you create this design? What are the most difficult features involved in the design? Take a few minutes to consider a modeling strategy and do preliminary planning by sketching on a piece of paper. You are also encouraged to create the design on your own prior to following through the tutorial.

Modeling Strategy

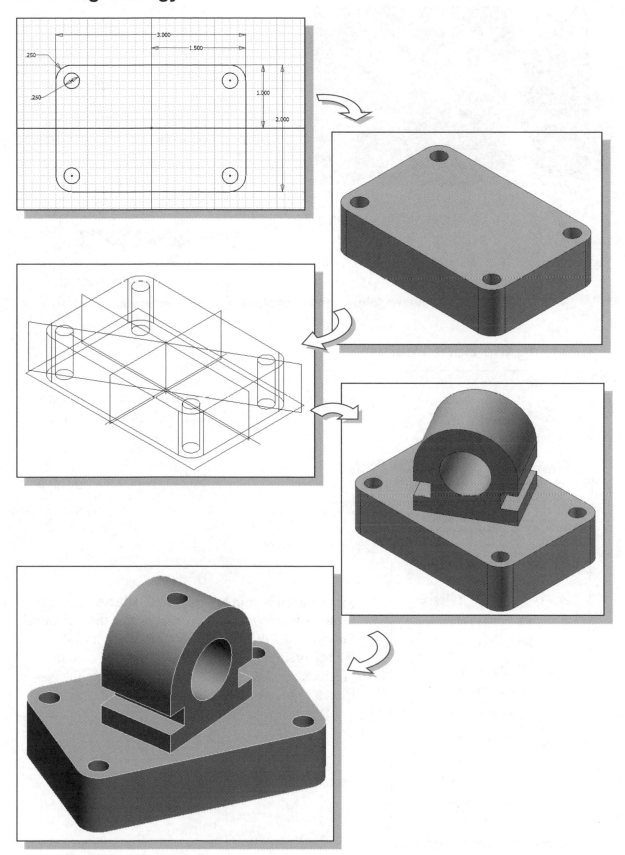

Starting Autodesk Fusion 360

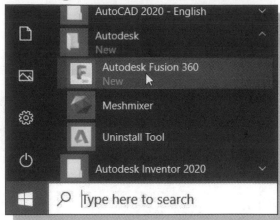

1. Select the **Autodesk Fusion 360** option on the *Start* menu or select the **Autodesk Fusion 360** icon on the desktop to start Autodesk Fusion 360.

2. In the *Sign In* dialog box, log in with your email or username.

3. On your own, set the modeling units to **Inches** in the *Browser* area.

Creating the Base Feature

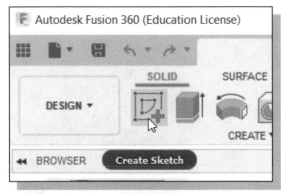

1. Activate the **Create Sketch** icon with a single click of the left-mouse-button.

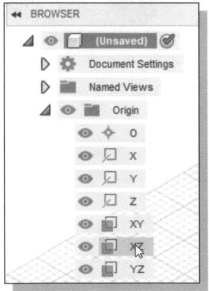

2. Move the cursor on top of the **XZ Plane**, inside the *browser* window as shown, and notice that Autodesk Fusion 360 will automatically highlight and show the orientation of the corresponding plane in the graphics window. Left-click once to select the XZ Plane as the sketching plane.

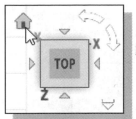

3. Single left-click to activate the **Home View** option as shown. The view will be adjusted back to the default *isometric view*.

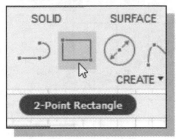

4. Select the **2-point rectangle** command by clicking once with the left-mouse-button on the icon in the *Sketch* toolbar.

5. Create a rectangle of arbitrary size with the center point near the center of the rectangle as shown.

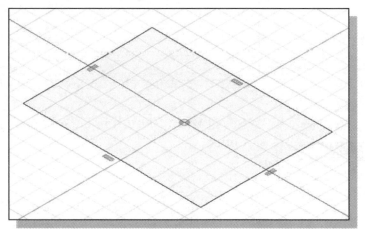

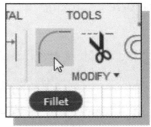

6. Click on the **Fillet** icon in the *Create Panel*.

7. The *2D Fillet* radius dialog box appears on the screen. Use the default radius value and create four rounded corners of the rectangle.

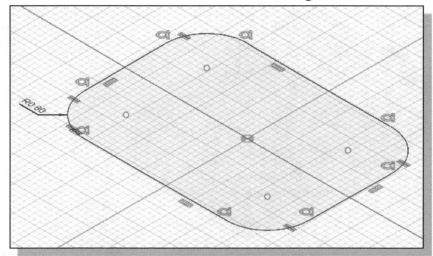

8. On your own, create four circles of the same diameter and with the centers aligned to the centers of the arcs. Also create and modify the six dimensions, 3x2 overall dimensions and radius/diameter 0.25, as shown in the figure.

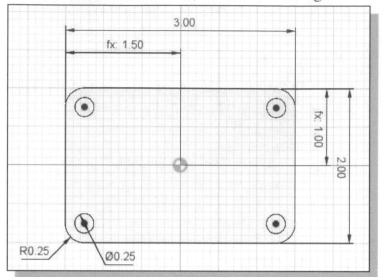

9. Select **Stop Sketch** in the *Ribbon* toolbar to end the Sketch option.

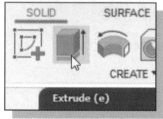

10. In the *3D Model* toolbar, select the **Extrude** command by clicking the left-mouse-button on the icon.

11. Select the inside region of the 2D sketch to create a profile as shown.

12. In the *Extrude* pop-up window, enter **0.75** as the extrusion distance and create the solid feature as shown.

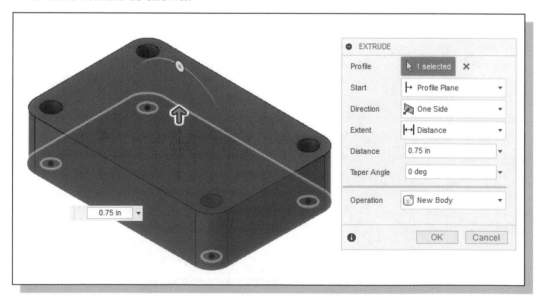

Create an Angled Work Plane

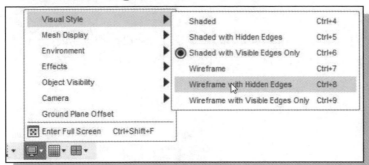

1. Select **Wireframe with Hidden edges** display mode under the *Visual Style* option as shown.

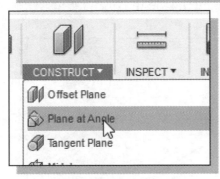

2. In the *Construct* panel, select the **Plane at Angle** command by left-clicking the icon.

   - Autodesk Fusion 360 expects us to select an existing line to be used as a reference to create the new work plane.

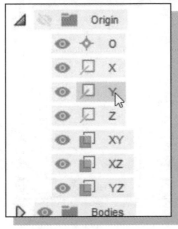

3. Inside the *browser* window, left-click once to select the **Y Axis** as the second reference of the new work plane.

4. In the *Angle* pop-up window, enter **30** as the rotation angle for the new work plane.

❖ Note that the *angle* is measured relative to the default reference plane, XY Plane.

5. Click on the **OK** button to accept the setting and create the work plane.

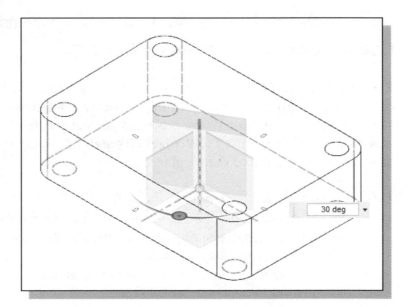

Create a 2D Sketch on the Work Plane

1. In the *Sketch* toolbar select the **Create Sketch** command by left-clicking once on the icon.

2. Pick the **work plane** by clicking the work plane name inside the *browser* as shown below.

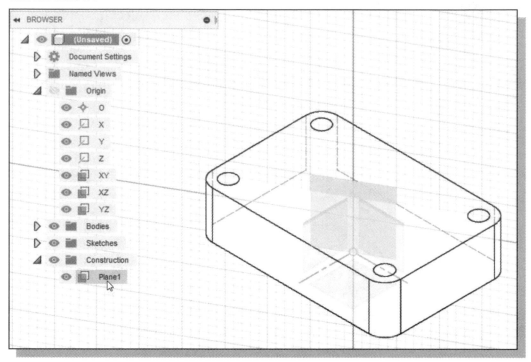

Use the Projected Geometry Option

Projected geometry is another type of *reference geometry*. The **Project Geometry** tool can be used to project geometry from previously defined sketches or features onto the sketch plane. The position of the projected geometry is fixed to the feature from which it was projected. We can use the **Project Geometry** tool to project geometry from a sketch or feature onto the active sketch plane.

Typical uses of projected geometry include:
*   Project a silhouette of a 3D feature onto the sketch plane for use in a 2D profile.
*   Project the default center point onto the sketch plane to constrain a sketch to the origin of the coordinate system.
*   Project a sketch from a feature onto the sketch plane so that the projected sketch can be used to constrain a new sketch.

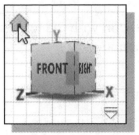

1. Single left-click to activate the **Home View** option as shown. The view will be adjusted back to the default *isometric view*.

2. Select the **Project Geometry** command in the *Create Panel*. The *Project Geometry* command allows us to project existing features to the active sketching plane.

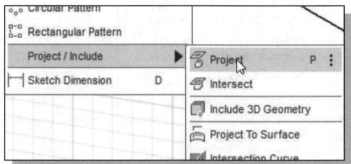

3. Select the top back edge of the base feature to create a projected line on the sketching plane.

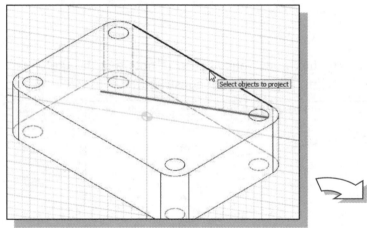

➢ The projected line will be used in the 2D profile of the next solid feature.

4. Click **OK** to accept the projected line.

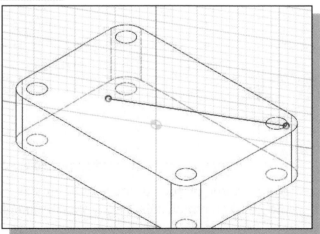

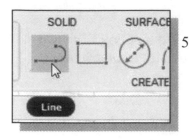

5. In the *Sketch* toolbar, click on the **Line** icon with the left-mouse-button to activate the Line command.

6. Create a rough sketch using the projected edge as the bottom line as shown in the figure. (Note that all edges are either horizontal or vertical.)

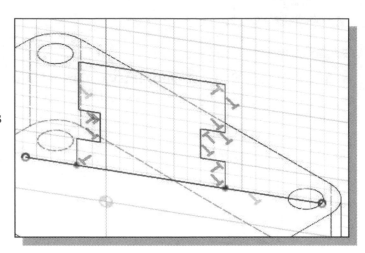

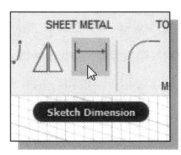

7. Activate the **Sketch Dimension** command in the *Create Panel* as shown.

8. On your own, create and modify the dimensions as shown; note that the sketch is symmetrical about the Y axis. (Hint: use the **Equal** constraint.)

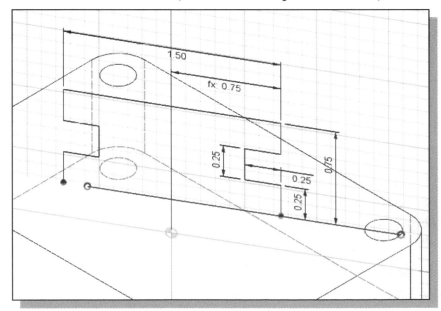

➢ Note that the projected line is used only as a reference; the gap at the bottom of the 2D sketch indicates the 2D sketch cannot be used as a profile yet.

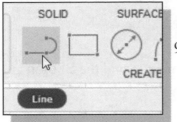

9. Select the **Line** command in the *2D Create Panel*.

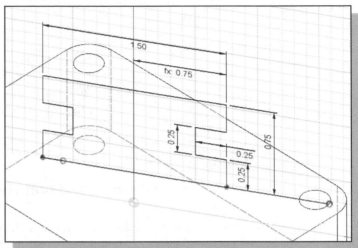

10. Create a line connecting the **bottom** two corners of the 2D sketch as shown in the figure.

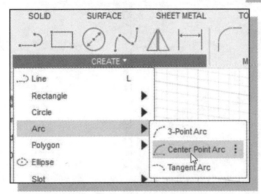

11. Select the **Center Point Arc** option in the *2D Create Panel* as shown.

12. On your own, create the arc so that the center point is aligned to the mid-point of the top edge.

13. On your own, add a **0.75** circle, apply collinear/equal length constraints, and complete the 2D sketch as shown in the figure.

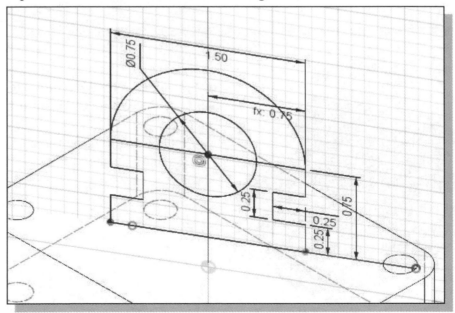

Complete the Solid Feature

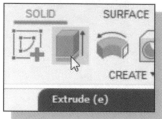

1. Select **Stop Sketch** in the *Ribbon* toolbar to end the Sketch option.

2. In the *Create* toolbar panel, select the **Extrude** command by clicking the left-mouse-button on the icon.

3. Select the two inside regions of the 2D sketch to create a profile as shown.

4. Set the extrusion direction option to **Symmetric – Whole Length** as shown.

5. In the *Extrude* pop-up window, enter **1.0** as the extrusion distance.

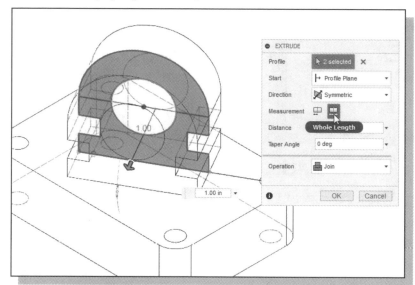

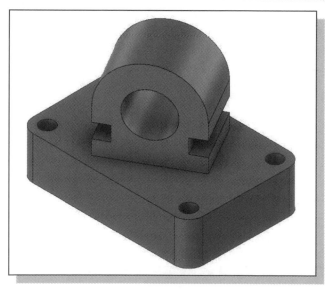

6. Click on the **OK** button to proceed with creating the feature.

7. On your own, set the display to the standard shaded mode.

Create an Offset Work Plane

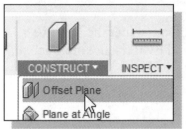

1. In the *Construct* panel, select the **Offset Plane** command by left-clicking the icon.

❖ Autodesk Fusion 360 expects us to select any existing geometry, which will be used as a reference to create the new work plane.

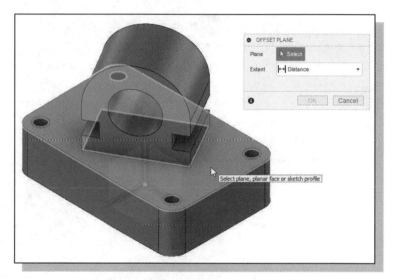

2. Inside the graphics window, select the **top plane** of the base feature as the reference of the new work plane.

3. Set the value in the *Offset* pop-up window to **0.75** as the offset distance for the new work plane.

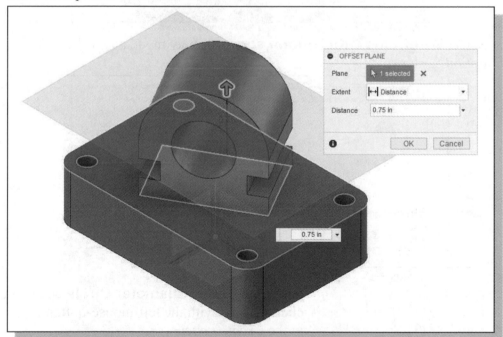

4. Click the **OK** button to accept the setting and create the reference plane.

Create another Cut Feature Using the Work Plane

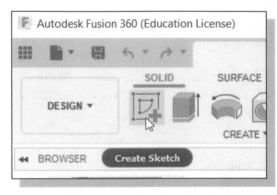

1. In the *Sketch* toolbar, select the **Create Sketch** command by left-clicking once on the icon.

2. Pick the work plane we just created by clicking on the work plane as shown below.

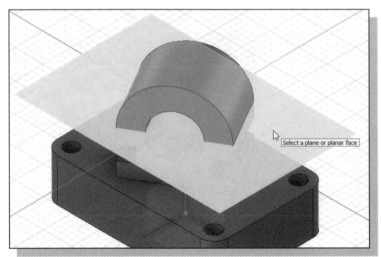

3. Select **Wireframe with Hidden edges** display mode under the *Visual Style* option as shown.

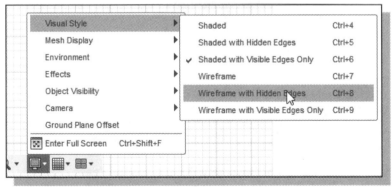

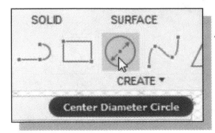

4. Select the **Center Diameter Circle** command by clicking once with the left-mouse-button on the icon in the *Create Panel*.

5. On your own, create a **circle** with the center point aligned to the **Origin** as shown in the figure below.

6. Using the *Sketch Dimension* command, set the circle to **Ø0.25**.

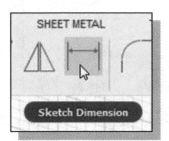

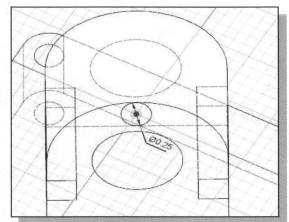

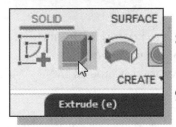

7. Click on the **Stop Sketch** icon to exit the 2D sketch mode.

8. In the *Create panel*, select the **Extrude** command by left-clicking once on the icon.

9. On your own, complete the **cut** feature as shown.

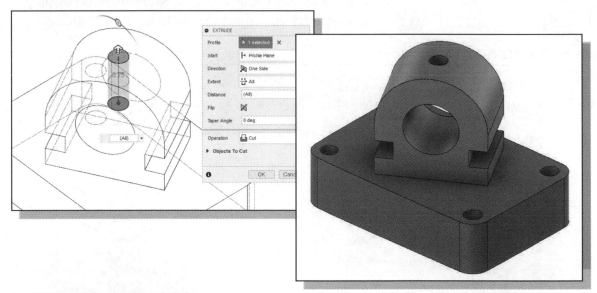

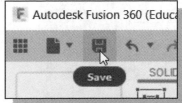

10. Save the design as ***Rod-Guide*** in the **Chapter 9** folder.

Start a New 2D Drawing

1. In the *File* pull down menu, select **New Drawing → From Design** to start a new drawing.

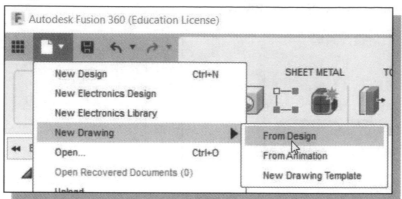

2. In the **Create Drawing** dialog box, set the Standard to ASME, units to in and sheet size to B (17in x 11 in) as shown and click OK.

➢ Note that a new tab appears in the Fusion 360 window. We can switch between the solid model and the drawing by clicking the corresponding tabs in the ribbon toolbar area.

❖ In the graphics window, Autodesk Fusion 360 displays a default drawing sheet that includes a title block.

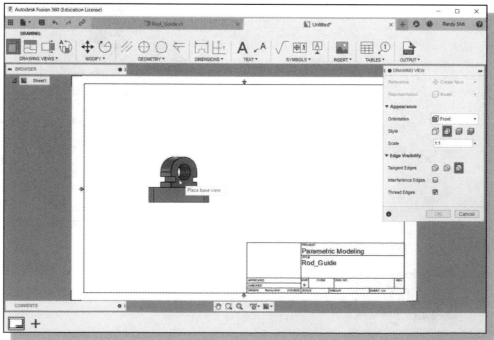

Add a Base View

In Autodesk Fusion 360 *Drawing Mode*, the first drawing view we create is called a **base view**. A *base view* is the primary view in the drawing; other views can be derived from this view. When creating a *base view*, Autodesk Fusion 360 allows us to specify the view to be shown. By default, Autodesk Fusion 360 will treat the *world XY plane* as the front view of the solid model. Note that there can be more than one *base view* in a drawing.

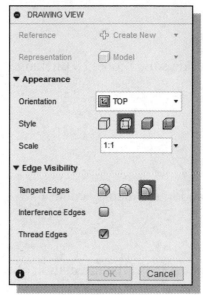

1. In the *Drawing View* dialog box, set *Orientation* to **Top** view, set the **Scale** to **1 : 1** and *Style* to **Hidden Line** as shown in the figure.

2. Confirm the *Tangent Edges* option is set to **off** as shown.

3. Inside the graphics window, place the **base view** near the upper left corner of the graphics window as shown below. (If necessary, drag the *Drawing View* dialog box to another location on the screen.)

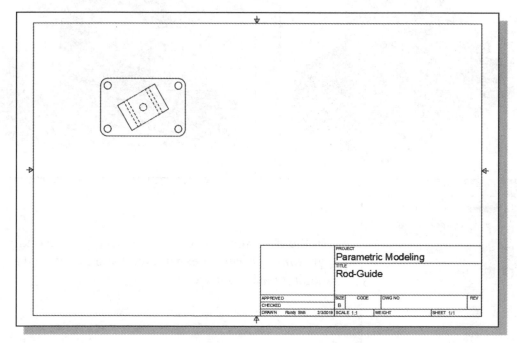

4. Click the [**OK**] key once to exit the *Drawing View* dialog box.

Create an Auxiliary View

In Autodesk Fusion 360 *Drawing Mode*, the **Projected View** command is used to create standard orthographic views such as the *top* view, *front* view or *isometric* view. For non-standard views, we will first create a new **Named View** in the 3D modeling mode. Note that *Auxiliary views* are created still using orthographic projections. Generally, orthographic projections are aligned to the base view and should inherit the base view's scale and display settings.

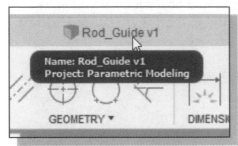

1. Switch to the Rod-guide solid model by clicking on the file tab as shown.

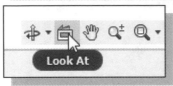

2. Activate the **Look At** command in the display control panel as shown.

3. Select the front face of the upper section of the 3D model to set the orientation of the view angle as shown.

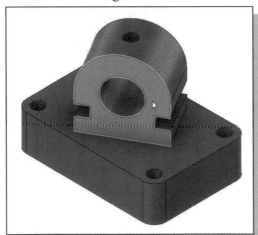
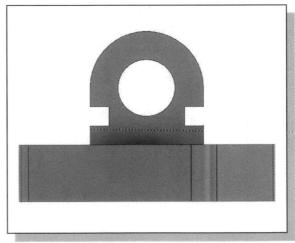

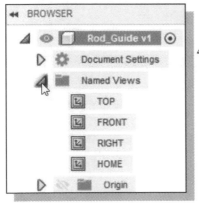

4. In the *Browser* area, expand the triangular icon in front of **Named Views** to expand the list and notice the default basic four views.

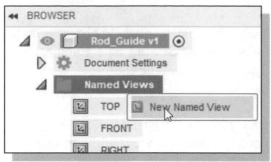

5. **Right-mouse-click** on *Named Views* to bring up the *option menu*.

6. Select **New Named View** option as shown.

7. Click on the newly created view and rename it to **Auxiliary** as shown.

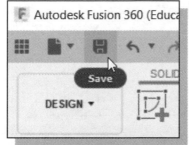

8. Click **Save** to accept the new changes and upload the design to the cloud server.

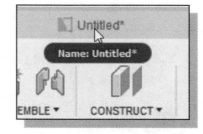

9. Switch to the Rod-guide drawing by clicking on the file tab as shown.

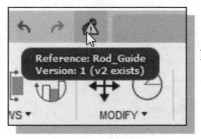

10. Pick the front edge of the upper section of the model as shown.

11. Click on **Base View** in the *Drawing Views* panel to create a new base view.

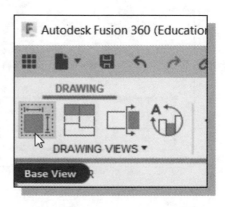

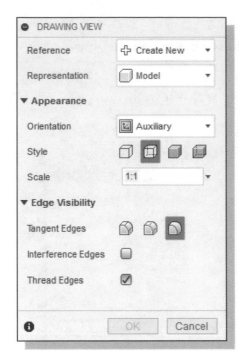

12. In the *Drawing View* dialog box, set *Orientation* to **Auxiliary** view, set the **Scale** to **1 : 1** and *Style* to **Hidden Line** as shown in the figure

13. Move the cursor below the base view and select a location to position the new view of the model as shown. Click **OK** to create the view.

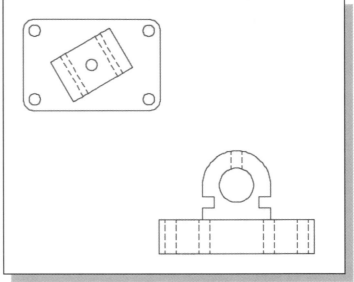

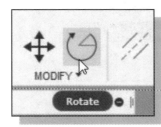

14. In the *Modify* panel, activate the **Rotate** command as shown.

15. Select the **Auxiliary view** as shown.

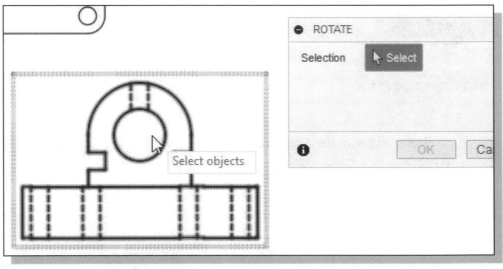

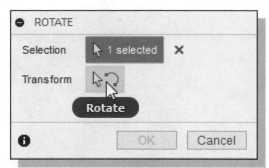

16. In the *Rotate* dialog box, activate the **Transform** option by clicking on the button as shown.

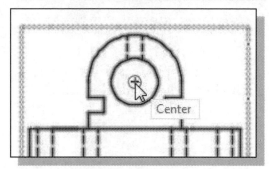

17. Select the **Center point** in the view as the reference point as shown.

18. Move the cursor to adjust the rotation angle; set the rotation angle to **30 degrees** as shown.

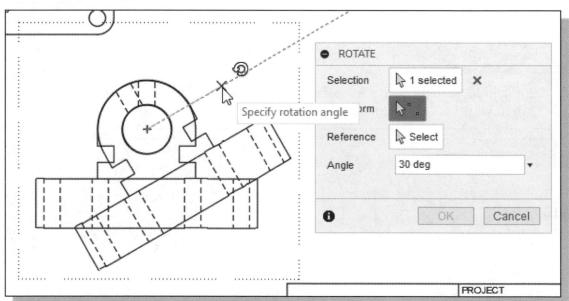

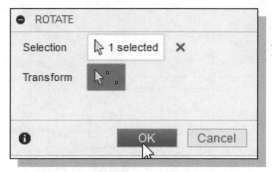

19. Click **OK** to end the *Rotate* command.

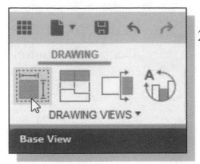

20. Click on **Base View** in the *Drawing Views* panel to create another base view.

21. In the *Drawing View* dialog box, set orientation to NE **Isometric**, **Scale 1 : 1** and **Hidden Line Removed** as shown in the figure.

22. Position the *isometric* view toward the right side of the title block as shown below. Click OK to create the view.

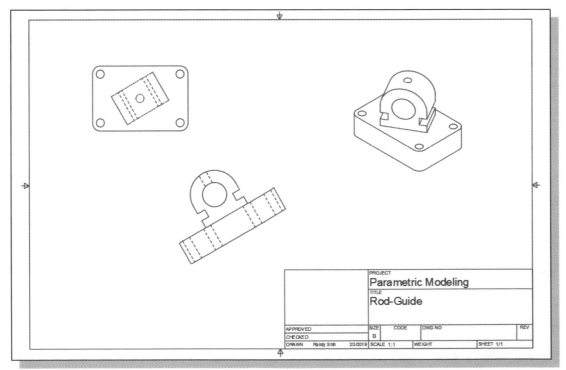

Adjust the View Scale

1. **Double-click** on the top view to enter the *Drawing View property* option.

2. Inside the *Drawing View* dialog box set the *Scale* to **3:2** as shown in the figure.

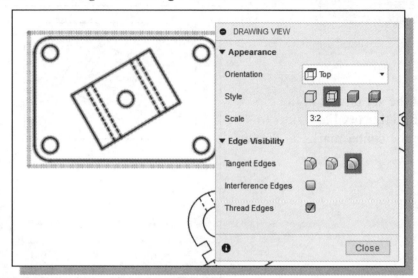

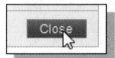   3. Click on the **Close** button to accept the settings.

4. On your own, repeat the above step and set the scale for the auxiliary also to 3:2. Reposition the views so that the drawing appears roughly as shown.

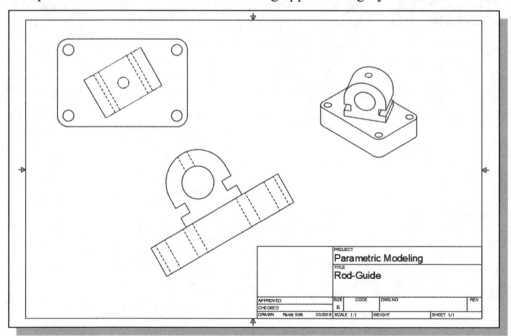

❖ Note that there are three base views currently in the 2D Drawing; the view properties are not associated together.

Aligning the top View and the Auxiliary View

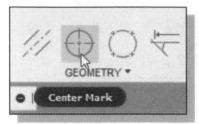

1. Click on the **Center Mark** button in the *Centerlines* panel as shown.

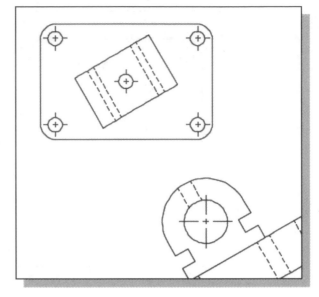

2. Select the six circles in the top view and the auxiliary view to create the center marks as shown.

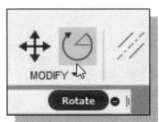

3. In the *Modify* panel, activate the **Rotate** command as shown.

4. On your own, rotate the center mark of the small circle in the top view and the center mark in the **auxiliary view** to 30 degrees as shown.

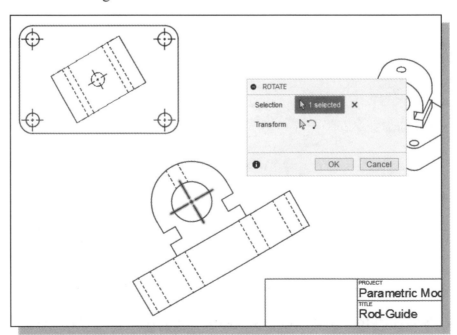

5. Click **OK** to accept the adjustment.

6. On your own, extend the center lines by using the grip point controls. Note that the lines aren't lined up.

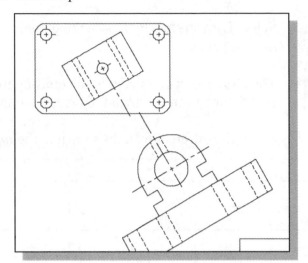

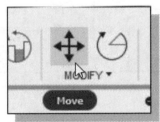

7. In the *Modify* panel, activate the **Move** command as shown.

8. Select the **auxiliary view** to move.

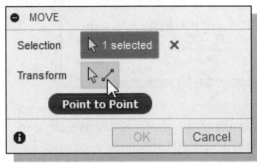

9. In the *Move* dialog box, activate the **Point to Point** option by clicking on the button as shown.

10. Click on the endpoint of the center line in the auxiliary and attach it to the corresponding endpoint in the top view as shown.

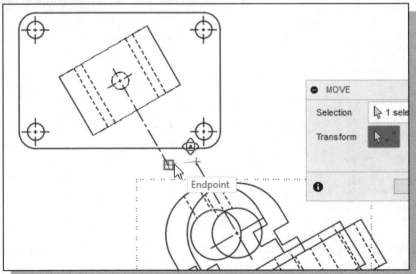

Add Dimensions

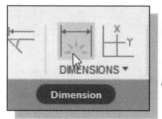

1. Select **Dimension** by left-clicking once in the *Dimensions* toolbar panel.

   • The **Dimension** command in the drafting module is the same as the dimension command in the modeling module.

2. Create the dimensions for the different features shown in the ***top*** view as shown.

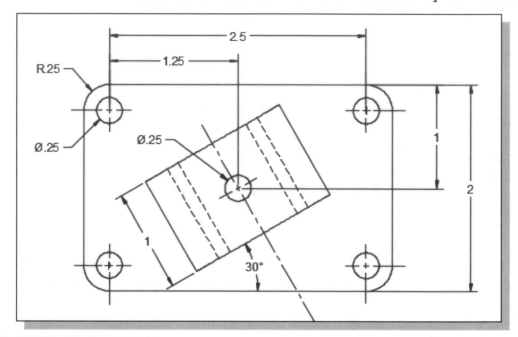

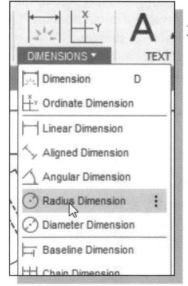

3. Note that different dimensioning options are available in the Dimensions panel for specific dimensions. For example, the *Radius Dimension* for the rounded corner as shown.

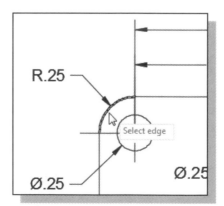

4. On your own, add additional centerlines to the auxiliary view and also extend the centerlines in both views.

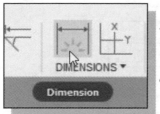

5. Select **Dimension** by left-clicking once in the *Dimensions* toolbar panel.

- The **Dimension** command in the drafting module is the same as the dimension command in the modeling module.

6. Create the dimensions for the different features shown in the *Auxiliary* view as shown.

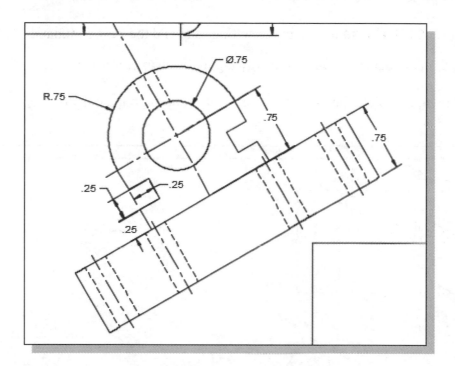

Complete the Drawing Sheet

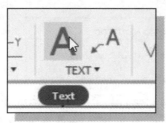

1. In the *Text Panel*, click on the **Text** button.

2. Pick a location that is inside the *DWG No* area and specify a text box region as shown.

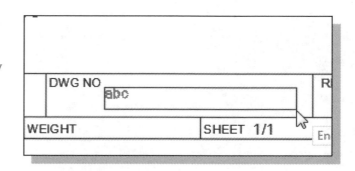

3. Enter **MECH 375-CH9-01** as the new *Drawing No* as shown.

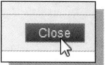

4. Click **Close** to accept the settings and end the *Text* command.

5. On your own, **Save** the drawing to the chapter 9 folder.

6. On your own, complete the multi-view drawing and print out a copy of the drawing as shown.

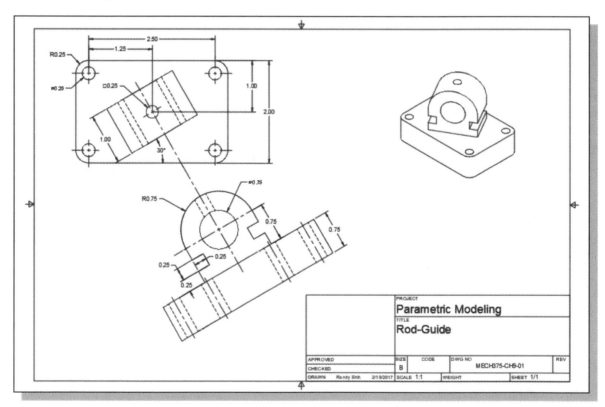

Review Questions:

1. What are the different types of work features available in Autodesk Fusion 360?

2. Why are work features important in parametric modeling?

3. Describe the purpose of auxiliary views in 2D drawings.

4. Describe the procedure to add an auxiliary view in the Drawing module.

5. Describe the different methods used to create centerlines in the chapter.

6. Can we change the *View Scale* of existing views? How?

7. What is the main difference between an auxiliary view and a projected view?

8. Describe the steps to change the *display style* of a drawing view.

9. Describe the difference between the centerlines created with the **Centerline Bisector** and the **Center Mark** commands.

Exercises: Create the Solid models and the associated 2D drawings and save the exercises in the Chapter 9 folder.

1. **Rod Slide** (Dimensions are in inches.)

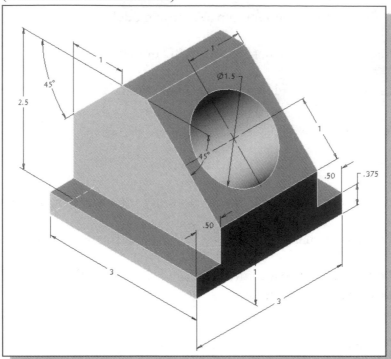

2. **Anchor Base** (Dimensions are in inches.)

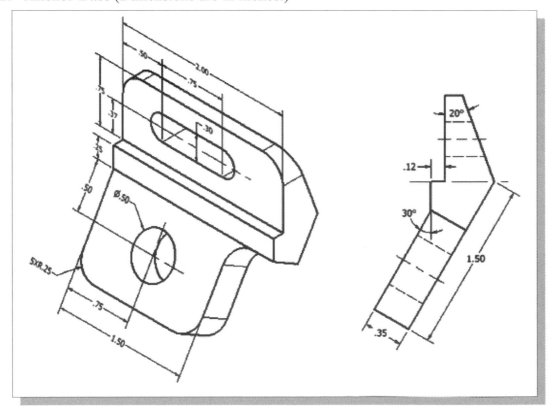

3. **Bevel Washer** (Dimensions are in inches.)

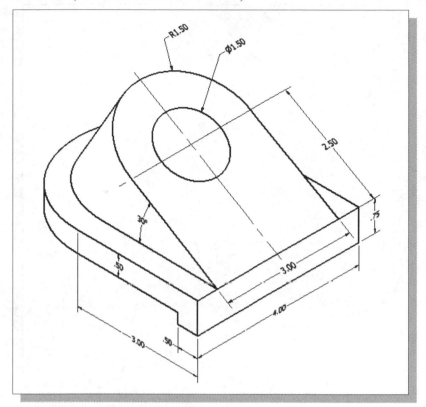

4. **Angle V-Block** (Dimensions are in inches.)

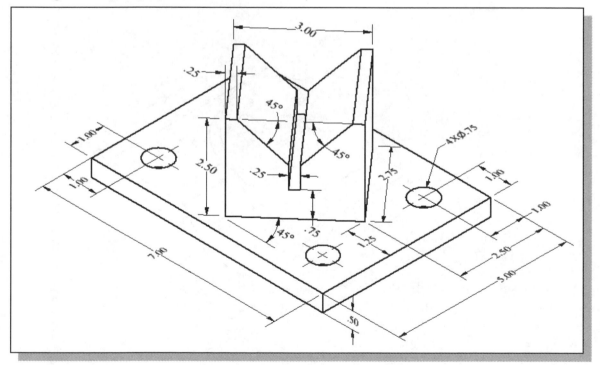

5. **Angle Support** (Dimensions are in millimeters.)

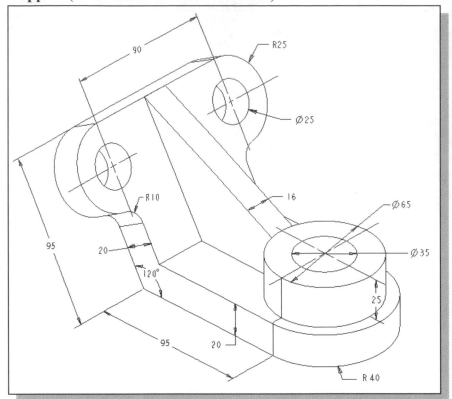

6. **Jig Base** (Dimensions are in millimeters.)

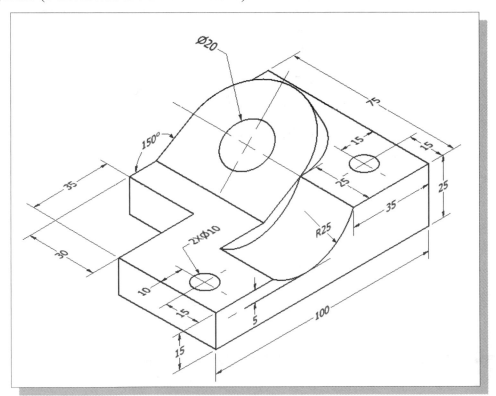

Chapter 10
Introduction to 3D Printing

Learning Objectives

♦ **Understand the History and Development of 3D Printing**

♦ **Be aware of the Primary types of 3D Printing Technologies**

♦ **Be able to identify the commonly used Filament types for Fused Filament Fabrication**

♦ **Understand the general procedure for 3D Printing**

What is 3D Printing?

3D Printing is a type of *Rapid Prototyping* (RP) method. Rapid prototyping refers to the techniques used to quickly fabricate a design to confirm/validate/improve conceptual design ideas. 3D printing is also known as **"Additive Manufacturing"** and construction of parts or assemblies is usually done by an addition of material in thin layers.

Prior to the 1980s, nearly all metalworking was produced by machining, fabrication, forming, and mold casting; the majority of these processes require the removal of material rather than adding it. In contrast to the *Additive Manufacturing* technology, the traditional manufacturing processes can be described as **Subtractive Manufacturing**. The term *Additive Manufacturing* gained wider acceptance in the 2010s. As the various additive processes continue to advance and become more mature, it is quite clear that they will compete with material removal as the main manufacturing process for many applications in the very near future.

The basic principle behind *3D printing* is that it is an additive process. 3D printing is a radically different manufacturing method based on advanced technology that create parts directly, by adding material layer by layer at the sub millimeter scale. One way to think about 3D Printing is the additive process is really performing "**2D printing over and over again**."

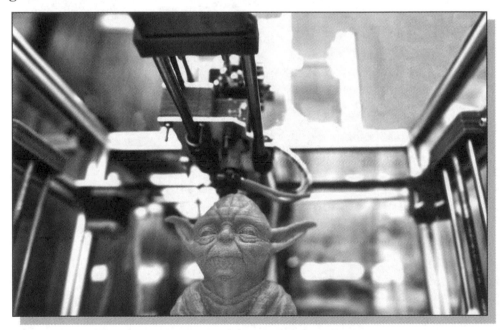

A number of limitations exist in the traditional manufacturing processes, which can be labor intensive, requiring expensive tooling, designing of fixtures, and the assembly of parts. The *3D printing* technology provides a way to create parts with complex geometric shapes quite easily using thin layers. The traditional *subtractive manufacturing* can also be quite wasteful as excess materials are cut and removed from large stock blocks, while the *3D printing* process basically uses only the material needed for the parts. *3D printing* is an enabling technology that encourages and drives innovation with unprecedented design freedom while being a tooling-less process that reduces costs and lead times. The relatively fast turnaround time also makes *3D printing* ideal for prototyping. Components with intricate geometry and complex features can also be designed specifically for *3D printing* to avoid complicated assembly requirements. *3D printing* is also an energy efficient technology that can provide better environmental friendliness in terms of the manufacturing process itself and the type of materials used for the product. There are quite a few different techniques to 3D print an object. *3D Printing* brings together two fundamental innovations: the manipulation of objects in the digital format and the manufacturing of objects by addition of material in thin layers.

The term **3D-printing** originally referred only to the smaller 3D printers with moveable print heads similar to an inkjet printer. Today, the term **3D-printing** is used interchangeably with **Additive Manufacturing**, as both refer to the technology of creating parts through the process of adding/forming thin layers of materials.

Development of 3D Printing Technologies

The earliest 3D printing technology was first invented in the 1980s; at that time it was generally called **Rapid Prototyping** (**RP**) technology. This is because the process was originally conceived as a fast and time-effective method for creating prototypes for product development in industry. In 1981, Dr. Hideo Kodama of *Nagoya Municipal Industrial Research Institute* invented two methods of creating three-dimensional plastic models with photo-hardening polymer through the use of a UV Laser. In 1986, the first US patent for **stereolithography** apparatus (**SLA**) was issued to Charles Hull, who first invented his SLA machine in 1983. Chuck Hull went on to co-found *3D Systems Corporation*, which is one of the largest companies in the 3D printing sector today. Chuck Hull also designed the **STL** (**ST**ereo**L**ithography) file format, which is widely used by 3D printing software performing the digital slicing and infill strategies common to the additive manufacturing processes. The first available commercial RP system, the **SLA-1** by *3D Systems* (as shown in the figure below), was made available in 1987.

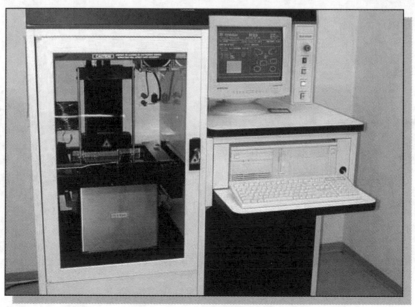

The 1980s also mark the birth of many RP technologies worldwide. In 1989, Carl Deckard of *University of Texas* developed the **Selective Laser Sintering (SLS)** process. In 1989, Scott Crump, one of the founders of *Stratasys Inc.*, also created the **Fused Deposition Modeling (FDM)**. In Europe, Hans Langer started *EOS GmbH* in Germany; the company focuses on the **Laser Sintering (LS)** process. The *EOS Systems Corp.* also developed the **Direct Metal Laser Sintering (DMLS)** process. Today, *3D Systems*, *EOS* and *Stratasys* are still the main leaders in the *Additive Manufacturing* industry.

During the 1990s, the *3D printing* sector started to show signs of distinct diversification with two specific areas of emphasis which are much more clearly defined today. First, there was the high end of 3D printing, still very expensive systems, which were geared towards part production for relatively complex designs. For example, in 1995, *Sciaky Inc* developed an additive welding process based on its proprietary **Electron Beam Additive Manufacturing (EBAM)** technology. Many *RP* system companies, such as *Solidscape*, *ZCorporation*, *Arcam* and *Objet Geometries* were all launched in the 1990s. At the other end of the spectrum, some of the 3D printing system manufacturers started to develop smaller desktop systems in the 1990s.

The idea of creating low-cost desktop 3D printers also intrigued many technology professionals and hobby enthusiasts during the late 1990s. In 2004, a retired professor, Dr Adrian Bowyer (the person on the left in the photo below), started the **RepRap** (*Replication Rapid-Prototyper*) project of an open source, self-replicating 3D printer (**RepRap 1.0 - Darwin**). This sets the stage for what was to come in the following years. It was around 2007 that the open source *3D printing* movement started gaining visibility and momentum. In January of 2009, the first commercially available open source 3D printer, the **BFB RapMan** 3D printer, became available. *Makerbot Industries* also came out with their **Makerbot** 3D printer in April of 2009. Since then, a host of low-cost desktop 3D printers have emerged each year.

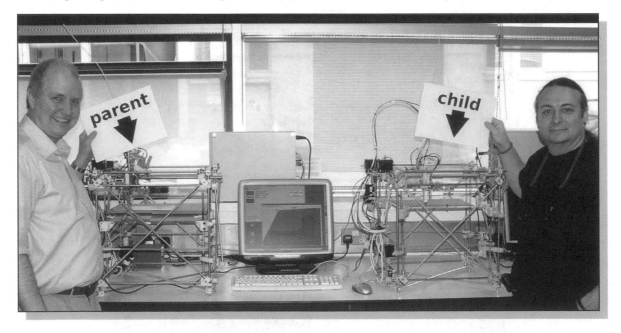

In the beginning of the 2010s, alternative 3D printing processes, such as those using **Polymer Resins**, became available at the desktop level of the market. The **B9 Creator**, using **Digital Light Processing (DLP)** technology, by *B9Creations*, came first in June of 2012, followed by the **Form 1** desktop printer by *Formlabs Inc.* Both 3D printers were launched via KickStarter's crowdfunding website and both enjoyed huge success. 2012 was also the year that many different mainstream media took note of the exciting 3D printing technology, which dramatically increased its visibility and awareness to the general public. 2013 was also a year of significant growth and consolidation; one of the most notable moves was the acquisition of Makerbot by Stratasys. In 2016, the new developments in 3D printing concentrated more on multi-color, multi-material using single or multiple extruders printers and new technologies to shorten the 3D printing time.

As a result of the market divergence, the price of desktop 3D printers continues to go down each year. Today, very capable fully assembled desktop 3D printers, such as Robo3D R1+ and Prusa I3 MK2, can be acquired for under $1000. Fully assembled smaller desktop 3D printers, such as XYZprinting's DA Vinci mini 3D Printer and M3D's Micro 3D, can be acquired for less than $350. Unassembled desktop 3D printer kits can even be acquired for under $200.

Another trend that happened in the 2010s is the availability of **3D printing Services**. 3D printing services are growing quite rapidly in the US. For example, many public libraries, especially in California, are now providing 3D printing services to the general public and **UPS** started its worldwide 3D printing services in May of 2016. This trend is spreading throughout the US, with many more companies planning to provide 3D printing services in the very near future. It is now quite feasible, and perhaps more economical, to 3D print designs without owning or ever touching a 3D printer, but understanding of the technology is still needed to increase productivity.

As the exponential adoption rate continues on all fronts, more and more technologies, materials, applications, and online services will continue to emerge. It is predicted that the development of 3D printing will continue in the years to come and 3D printing will eventually become the mainstream manufacturing method in industries and homes.

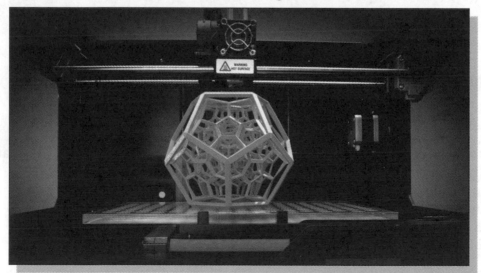

Primary Types of 3D Printing Processes

There are quite a few different techniques to 3D print an object. The different types of 3D printers each employ a different technology that processes different materials in different ways. For example, some 3D printers process powdered materials (nylon, plastic, ceramic and metal), which utilize a light/heat source to sinter/melt/fuse layers of the powder together in the defined shape. Others process polymer resin materials and again utilize a light/laser to solidify the resin in thin layers. **Stereolithography (SLA or SL), Fused Deposition Modeling (FDM or FFF)** and **Laser Sintering (LS or SLS)** represent the three primary types of 3D printing processes; the majority of the other 3D printing technologies are variations of the three main types.

Stereolithography

Stereolithography (**SLA** or **SL**) is widely recognized as the first 3D printing process; it was certainly the first to be commercialized. *SLA* is a laser-based process that works with photopolymer resins. The photopolymer resins react with the laser and cure to form a solid in a very precise way to produce very accurate parts. It is a complex process, but simply put, the photopolymer resin is held in a container with a movable platform inside. A laser beam is directed in the X-Y axes across the surface of the resin according to the 3D data supplied to the machine. The resin hardens precisely as the laser hits the designated area. Once the current layer is completed, the platform within the container drops down by a fraction (in the Z axis) and the subsequent layer is traced out by the laser. This 2D layer tracing continues until the entire object is completed.

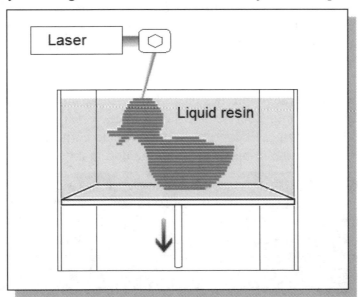

Because of the nature of the SLA process, support structures are needed for some parts, specifically those with overhangs or undercuts. These support structures need to be removed once the part is created. Many 3D printed objects using SLA need to be further cleaned and/or cured. Curing involves subjecting the part to intense light in an oven-like machine to fully harden the resin. SLA is generally accepted as being one of the most accurate 3D printing processes with excellent surface finish.

Fused Deposition Modeling (FDM) / Fused Filament Fabrication (FFF)

3D printing utilizing the extrusion of thermoplastic material is probably the most popular 3D printing process. The original name for the process is **Fused Deposition Modeling (FDM)**, which was developed in the early 1990s and is a trade name registered by *Stratasys*. However, a similar process, **Fused Filament Fabrication (FFF)**, has emerged since 2009. The majority of the desktop 3D printers, both open source and proprietary, utilize the FFF process, which is a more basic extrusion form of FDM.

The FDM and FFF processes work by melting plastic filament that is deposited, via a heated extruder, one layer at a time, onto a build platform according to the 3D data supplied to the 3D printer. Each layer hardens as it cools down and bonds to the previous layer.

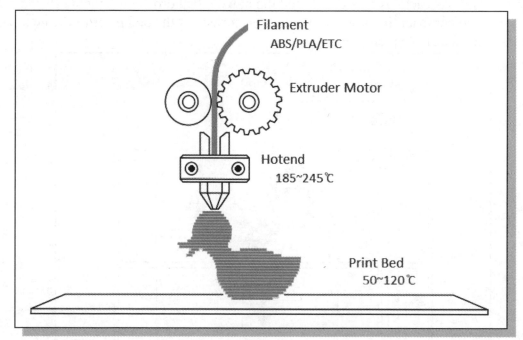

Stratasys has developed a range of proprietary industrial grade materials for its FDM process that are suitable for production applications. However, the most common materials for both FDM and FFF 3D printers are **ABS (Acrylonitrile Butadiene Styrene)** and **PLA (Polylactic Acid)**. The FDM and FFF processes require support structures for any applications with overhanging geometries. This generally entails a second, typically water-soluble or breakaway material, which allows support structures to be easily removed once the print is complete.

The FDM and FFF printing processes can be slow for large parts or parts with complex geometries. The layer to layer adhesion can also be a problem, resulting in parts that warp or separate easily. The surface finish of FDM and FFF printed parts might appear a bit rough as the thin layers are generally visible. To improve the appearance, several options are feasible, such as using acetone, sanding and/or spray paint.

Laser Sintering / Laser Melting

Laser Sintering (LS) or **Selective Laser Sintering (SLS)** creates tough and geometrically intricate parts using a high-powered CO_2 laser to fuse/sinter/melt powdered thermoplastics. The main advantage of SLS *3D printing* is that as a part is made, it remains encased in powder; this eliminates the need for support structures and allows for very complex 3D geometries to be 3D printed. SLS can be used to produce very strong parts as exceptional materials such as Nylon and metal powders are commonly used.

Laser sintering refers to a laser-based 3D printing process that works with powdered materials. The laser is traced across a powder bed of tightly compacted powdered material, according to the 3D data provided to the machine, in the X-Y axes. As the laser interacts with the powdered material it sinters and fuses the particles to each other forming a solid. As each layer is completed the powder bed drops incrementally and a roller is used to compact the powder over the top surface of the bed prior to the next pass of the laser for the subsequent layer.

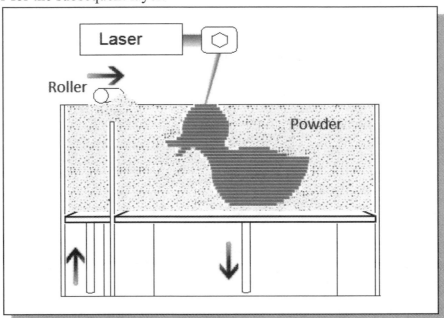

The build chamber is completely sealed as it is necessary to maintain a precise temperature during the process specific to the melting point of the powdered material of choice. One of the key advantages of this process is that the powder bed serves as an in-process support structure for overhangs and undercuts, and therefore complex shapes that could not be manufactured in any other way become possible with this process. Because of the high temperatures required for laser sintering, cooling can take a long time. Porosity is also a common issue with this process; additional metal infiltration processes may be required to improve mechanical characteristics.

Laser sintering can process plastic and metal materials, although metal sintering does require a much higher-powered laser and higher in-process temperatures. Parts produced with this process are much stronger than parts made with SLA or FDM, although generally the surface finish and accuracy is not as good.

Primary 3D Printing Materials for FDM and FFF

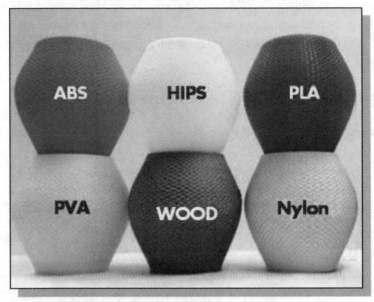

ABS (Acrylonitrile Butadiene Styrene)

ABS is a popular choice for 3D printing. It is a strong thermoplastic that is among the most widely used plastics. It is tough with mild flexibility, making it more durable to stress and has a higher heat resistance of up to 200 degrees Fahrenheit. However, this material has a tendency to shrink, which can affect the accuracy of designs. ABS has a pretty high melting point, and can experience warping if cooled while printing. Because of this, ABS objects are printed typically on a heated surface. ABS also requires ventilation when in use, as the fumes can be unpleasant. The aforementioned factors make ABS printing difficult for hobbyist printers, though it's the preferred material for professional applications.

PLA (Polylactic Acid)

PLA is a staple and it is becoming one of the most popular choices for 3D printing with good reasons. In addition to the fact that it is a biodegradable thermoplastic derived from renewable resources such as corn starch, tapioca roots, chips or starch, or sugarcane, *PLA* is a very rigid material that is easy to use for 3D printing and it is able to withstand a good amount of impact and weight. It also has a glossier finish than ABS and in most scenarios PLA is the preferred material for 3D printing large objects. The main disadvantage of PLA is it's not as heat resistance as ABS; it should not be placed in environments that exceed 140 degrees Fahrenheit.

Flexible (Thermoplastic Elastomer)

Flexible material is for applications that require incredible rubbery flex in their applications. Flexible filament goes beyond bending; it is more like rubber. When it comes to Flexible filament, it's all about finding a balance between flexibility (softness) and printability. This softness is sometimes indicated with a *Shore* value (like 85A or 60D). A higher Shore value means less flexibility. Harder filaments (less flexible) are easier to 3D print when compared to softer, more flexible filaments.

PETG (Polyethylene Terephthalate)
PETG is a material that is similar to *PLA*, with more attractive characteristics such as being generally a tougher and denser material with good heat resistance of up to 190 degrees Fahrenheit. It is reported to have the strength of *ABS*, while printing as easy as *PLA*.

HIPS (High Impact Polystyrene) and **PVA (Polyvinyl Alcohol)**
HIPS and *PVA* are relatively new materials that are growing in popularity for their dissolvable properties. They are used for creating support material. Their ability to dissolve in certain liquids means that they can be easily removed. These materials can be hard to print with, because they don't stick well to the build plates. It is also important not to print *PVA* at too high a temperature, as it can turn into tar and jam the extruder.

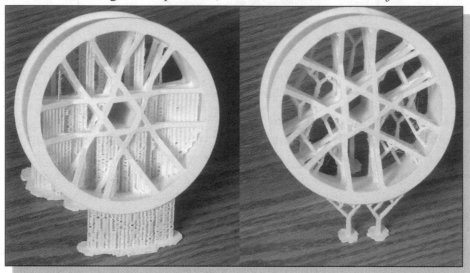

Wood Fiber
Wood Fiber filament contains a mixture of recycled wood with a binding polymer. Thus, a 3D printed object can look and smell like real wood. Due to its wooden nature, it's difficult to tell that the object is 3D printed. Using Wood filament is similar to using a thermoplastic filament like ABS or PLA. However, a 3D object having a wooden-like appearance can be created with this material.

From 3D Model to 3D Printed Part

To create a 3D printed part, it all starts with making a virtual design of the object. This virtual design may be created with a computer-aided design (CAD) package, via a 3D scanner, or by a digital camera and photogrammetry software. 3D scanning and photogrammetry software can process the collected digital data on the shape and appearance of a real object and create a digital 3D model. The 3D virtual design can generally be modified with 3D CAD packages, allowing verification of the virtual design before it is 3D printed.

Once the virtual design is verified, the 3D data will then be transferred to the 3D printing software. There is a multitude of file formats that 3D printing software supports. However, the most popular are the STL file format and the OBJ file format. The STL file format is the most commonly used file format for 3D printing. Most CAD software has the capability of exporting models in the STL format. The STL file contains only the surface geometry of the modeled object. The OBJ file format is considered to be more complex than the STL file format as it is capable of displaying texture, color and other attributes of the three-dimensional object. However, the STL file format holds the top spot for 3D printing, as this file format is simpler to use and most CAD packages work better with STL files than OBJ files.

Once the 3D data of the virtual design is transferred into the 3D printing software, further examination and/or repair can be performed if necessary. The 3D printing software will also process the imported 3D data by the special software known as a **Slicer**, which converts the model into a series of thin layers and produces a G-code file containing instructions tailored to a specific type of 3D printer. G-code is the common name for the most widely used numerical control (NC) programming language. It is used mainly in computer-aided manufacturing to control automated machine tools. The generated G-code file can be sent to the 3D printer and create the 3D printed part.

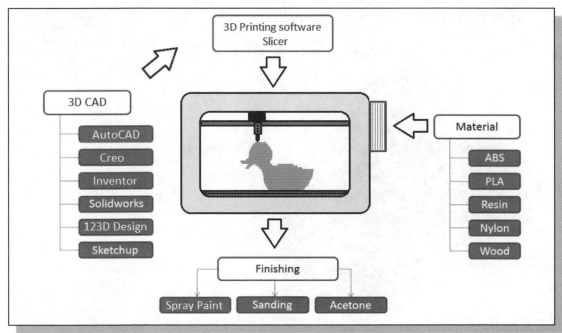

Starting Autodesk Fusion 360

1. Select the **Autodesk Fusion 360** option on the *Start* menu or select the **Autodesk Fusion 360** icon on the desktop to start Autodesk Fusion 360.

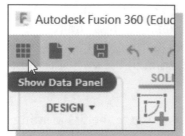

2. Click on the **Show Data Panel** icon as shown.

3. Open the *Parametric Modeling* project and double click on the **Chapter 9** folder.

4. Double click on the **Rod-Guide** model to open the model as shown.

5. On your own, change the *Units* option to **millimeter**. Note that the majority of the 3D printer settings are measured in millimeters, such as filament diameter, layer height and extruder size.

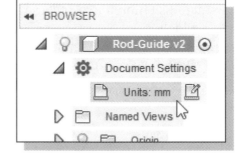

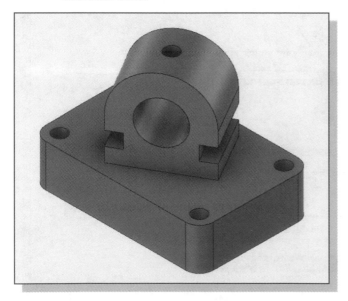

Export the Design as an STL file

3D printers will generally accept a 3D model with the STL or OBJ file formats. FUSION 360 offers two options to saving the 3D model in STL format: 1. Using the **[Files →3D Print]** option or 2. Using the FUSION 360 **[Make 3D Print]** command available for subscribed users.

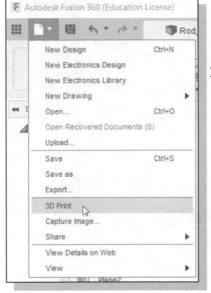

1. In the *Make Toolbar Panel,* click on the make icon to display the option list.

2. In the *3D Print* dialog box, set the *Refinement* to **High** and turn **off** the *Send to 3D Print Utility* option as shown.

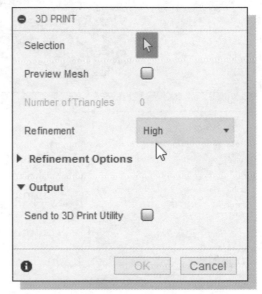

3. Fusion 360 expects us to select a model to be processed; select the *Rod-Guide* design as shown.

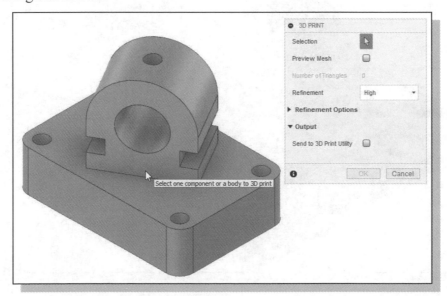

4. In the *3D Print* dialog box, click the **Mesh preview** option to view the 3D model. The STL file will contain 3566 triangles representing the solid model.

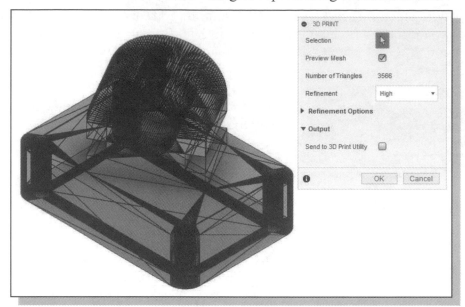

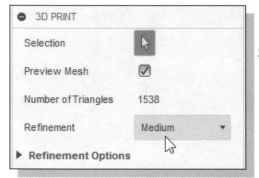

5. Uncheck the **Mesh preview** option and adjust the refinement to **Medium** and notice the number of triangles drops to 1538. The fewer number of triangles indicates a rougher representation of the model.

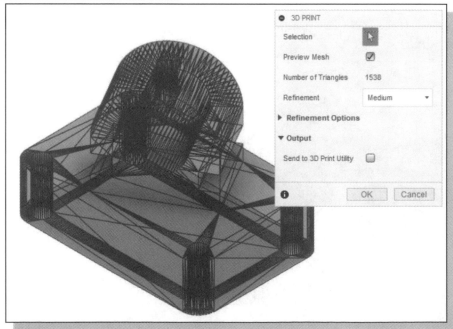

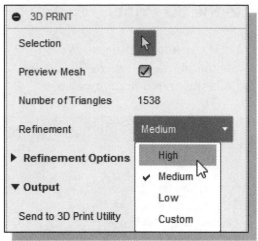

6. In the *3D Print* dialog box, set the *Refinement* to **High** as shown.

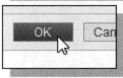

7. Click **OK** to accept the settings for exporting the 3D model.

8. On your own, select an *output folder*. This is the location for the **STL** file to be saved.

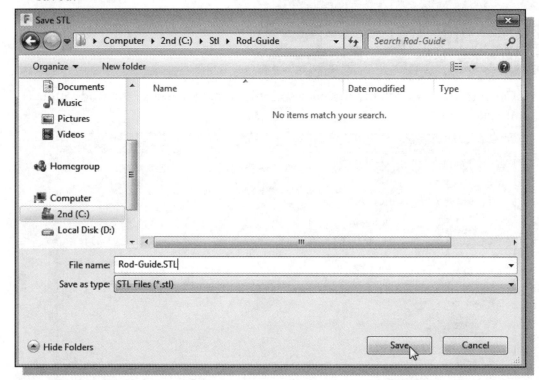

9. Enter **Rod-Guide.STL** as the filename as shown. Note that FUSION 360 automatically set the file format type to STL.

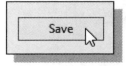

10. Click **Save** to proceed with saving the STL file.

Using the 3D Printing software to create the 3D Print

To 3D print the model, we will open the STL file in the 3D printing software. We will use **Matter Control** to demonstrate the procedure. Note that *Matter Control* (Freeware) supports quite a few desktop 3D printers. The procedures illustrated here are also applicable to other similar software.

1. Start the **Matter Control** software.

2. In the *toolbar area*, select **Open File** as shown.

3. On your own, switch to the saved STL file folder and select the Rod-Guide.STL file as shown.

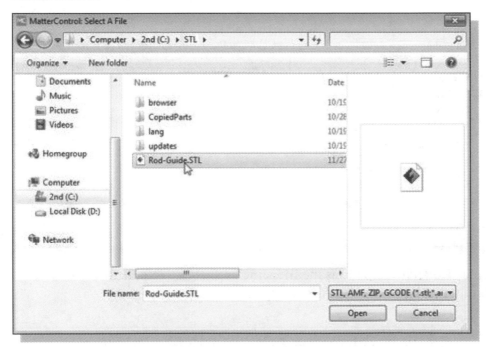

4. Click **Open** to import the STL file in to Matter Control.

5. Once the STL file is imported into the program, the STL model is displayed in the *View* window. Note that the model is imported with the incorrect orientation of the model; the bottom side of the model is not aligned to the print bed.

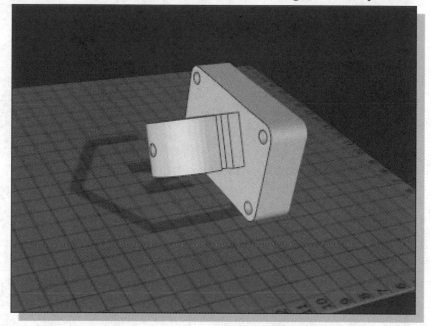

6. To adjust the display of the model, use the three mouse button to perform the Pan, Zoom and Rotate functions.

7. Note that the View Cube, located on the right side of the graphics window, is also available to control the viewing direction of the print bed.

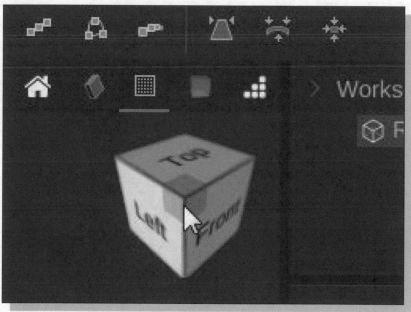

8. Click on the model, with the left mouse button, to enter the *Edit mode* and click & drag the model to reposition the model on the print-bed.

9. To rotate the mode, click on the **Rotate icon** to enter the *Rotate control*; note the associated dial allows more precise rotation.

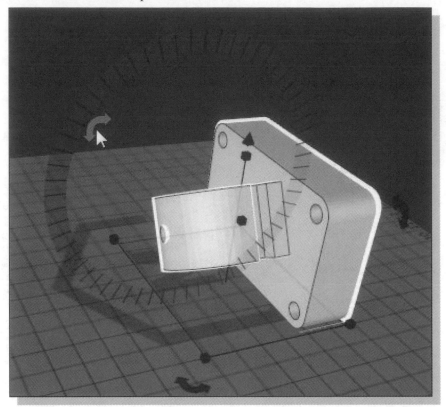

10. On your own, rotate the model so that it is setting vertically as shown.

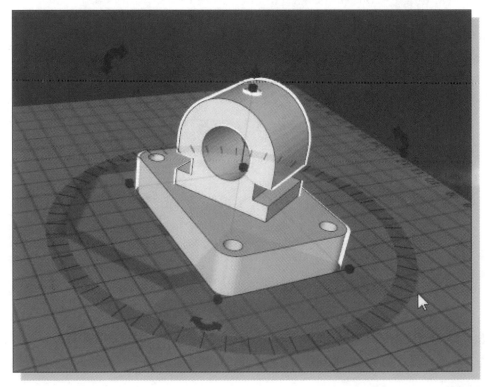

11. Note that additional *Editing tools*, such as **Scale** and **Mirror**, are also available through the right-mouse click on the model as shown.

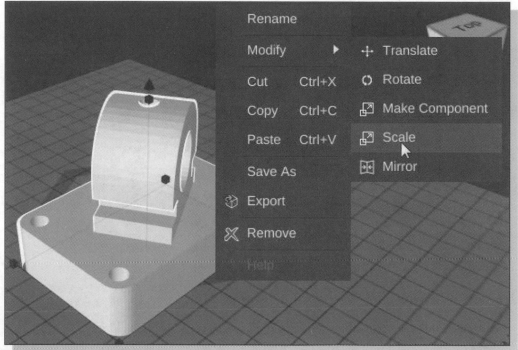

12. In the toolbar area, click **Lay Flat** to align the bottom surface of the model to the print bed.

13. Click **Slice Settings** to review the 3D Printing settings. Note that the **Control tab** contains commands to directly control the movements of the 3D printer.

14. Under the **Slice Settings tab,** different settings are available to adjust the 3D printing settings.

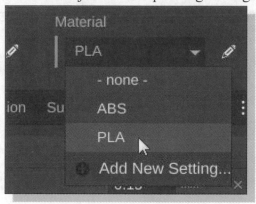

15. Under the **General** tab, a list of layer settings is available, such as the layer thickness, Top and Bottom Solid Layers thickness and the Infill type.

16. Under the **Speed** tab, the list of different speed settings on 3d printing is displayed and can be adjusted.

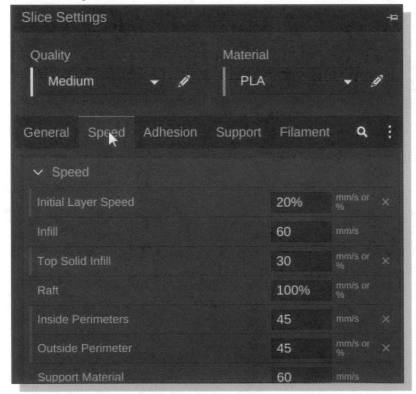

17. Under the **Adhesion** tab, the list of settings on improving adhesion on the first layer, such as Skirt, Raft and Brim, is available.

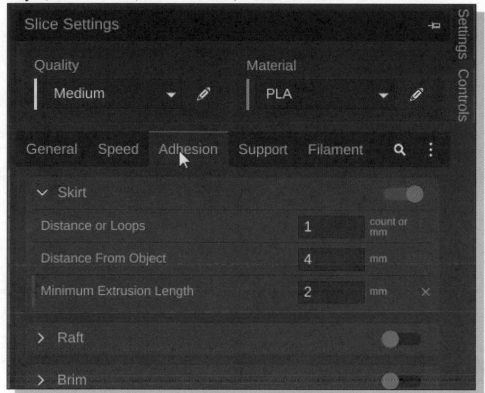

18. Under the **Support tab,** the adding support option can be turned on to *generate Support Material* as shown.

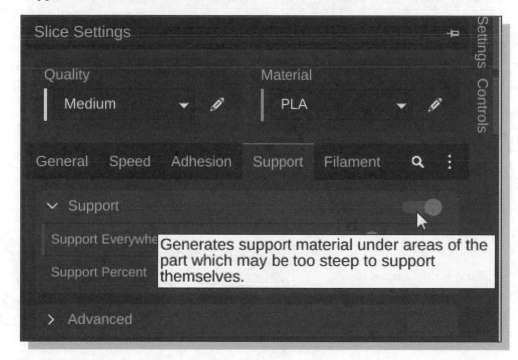

19. Switch to the **Filament tab** and in the **Material list** confirm/modify the filament properties, such as the diameter, to match the actual filament being used.

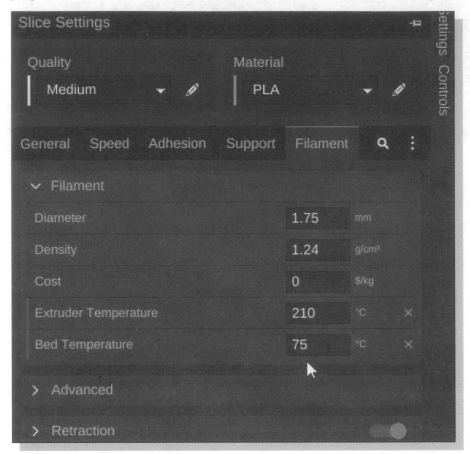

20. The temperature for the **Extruder** and the print bed can be adjusted on this page as well. Note that the temperatures are different based on the types of filaments, as well as the different brands and the type of printer bed. It is necessary to do some testing and/or experimenting when a new roll of filament is used.

21. In the toolbar area, click **Slice** to process the 3D model, which includes slicing and generating the associated G-code for the specific 3D printer.

22. Depending on the size and complexity of the design, it might take several minutes to complete the process.

23. On your own, drag the vertical slider to review the thin layers generated by the slicer.

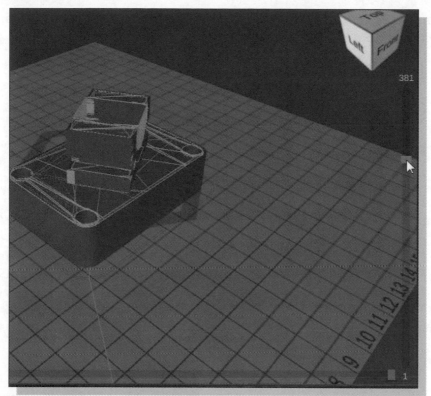

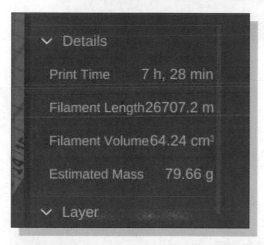

24. Note that with the current settings, it will take 7 hours and 28 minutes to complete the print using 26707.2 mm of filament and the volume of the printed part is 79.66gram.

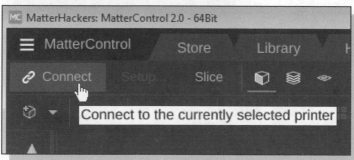

25. To start the 3D print, switch on the 3D printer and click Connect → Print to start the printing of the 3D model.

26. *Matter Control* will now begin printing.

Review Questions:

1. What is the main difference between the **Additive Manufacturing** and the traditional **Subtractive Manufacturing** technologies?

2. Which 3D printing process is recognized as the first 3D printing process?

3. Describe the general procedure to create a 3D printed part.

4. What are the three primary types of 3D printing processes?

5. Which 3D printing process is the most popular 3D printing process?

6. What is the main advantage of using PLA over ABS for FFF process?

7. Which are the most popular file formats for 3D printing?

8. What is the main function of a **Slicer** program?

Chapter 11
Symmetrical Features in Designs

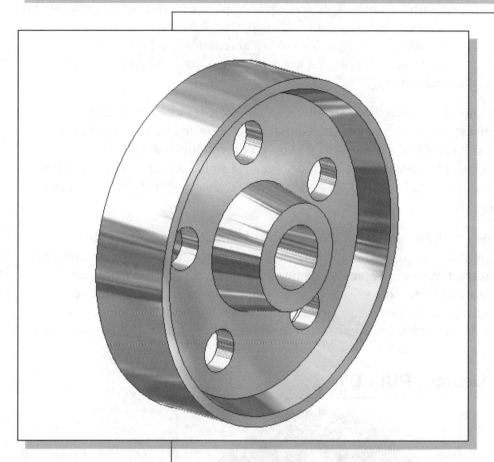

Learning Objectives

- Create Revolved Features
- Use the Mirror Feature Command
- Create Circular Patterns
- Create and Modify Diameter Dimensions
- Identify Symmetrical Features in Designs

Introduction

In parametric modeling, it is important to identify and determine the features that exist in the design. *Feature-based parametric modeling* enables us to build complex designs by working on smaller and simpler units. This approach simplifies the modeling process and allows us to concentrate on the characteristics of the design. Symmetry is an important characteristic that is often seen in designs. Symmetrical features can be easily accomplished by the assortment of tools that are available in feature-based modeling systems, such as Autodesk Fusion 360.

The modeling technique of extruding two-dimensional sketches along a straight line to form three-dimensional features, as illustrated in the previous chapters, is an effective way to construct solid models. For designs that involve cylindrical shapes, shapes that are symmetrical about an axis, revolving two-dimensional sketches about an axis can form the needed three-dimensional features. In solid modeling, this type of feature is called a *revolved feature*.

In Autodesk Fusion 360, besides using the **Revolve** command to create revolved features, several options are also available to handle symmetrical features. For example, we can create multiple identical copies of symmetrical features with the **Feature Pattern** command, or create mirror images of models using the **Mirror Feature** command. We can also use *construction geometry* to assist the construction of more complex features. In this lesson, the construction and modeling techniques of these more advanced options are illustrated.

A Revolved Design: PULLEY

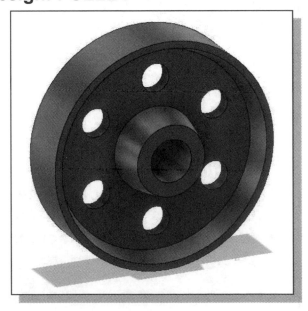

❖ Based on your knowledge of Autodesk Fusion 360, how many features would you use to create the design? Which feature would you choose as the **base feature** of the model? Identify the symmetrical features in the design and consider other possibilities in creating the design. You are encouraged to create the model on your own prior to following through the tutorial.

Modeling Strategy – A Revolved Design

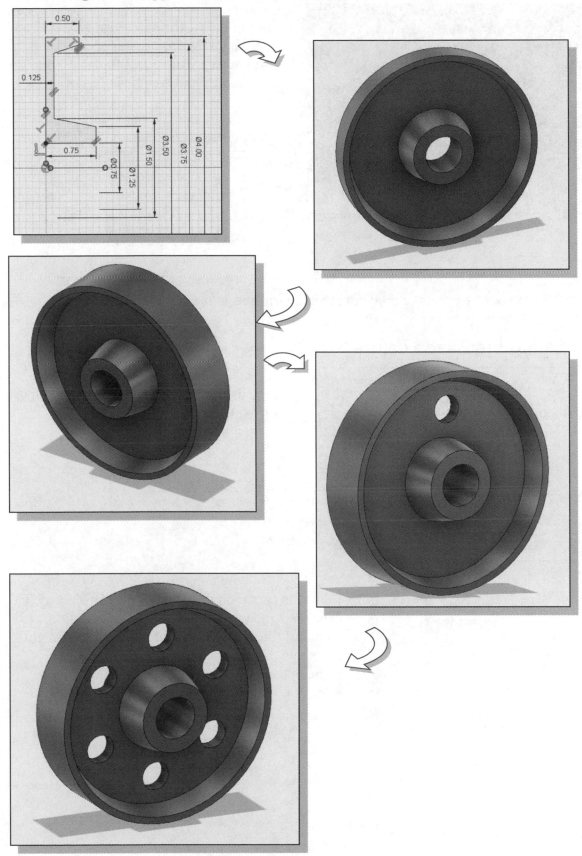

Starting Autodesk Fusion 360

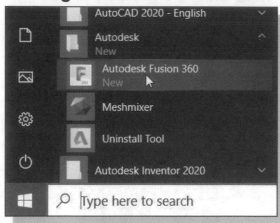

1. Select the **Autodesk Fusion 360** option on the *Start* menu or select the **Autodesk Fusion 360** icon on the desktop to start Autodesk Fusion 360.

2. In the *Sign In* dialog box, log in with your email or username.

3. On your own, set the modeling units to **Inches** in the *Browser* area.

Creating the Base Feature

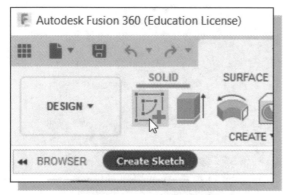

1. Activate the **Create Sketch** icon with a single click of the left-mouse-button.

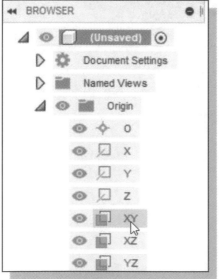

2. Move the cursor on top of the **XY Plane**, inside the *browser* window as shown, and notice that Autodesk Fusion 360 will automatically highlight and show the orientation of the corresponding plane in the graphics window. Left-click once to select the XY Plane as the sketching plane.

3. Select the **Project Geometry** command in the *Sketch* panel. The *Project Geometry* command allows us to project existing features to the active sketching plane.

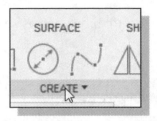

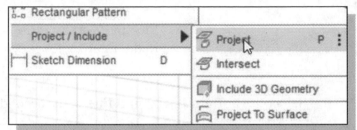

4. Inside the *browser* window, select the **X-axis** and **Y-axis** to project these entities onto the sketching plane. Click **OK** to accept the selection.

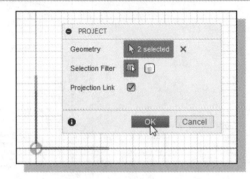

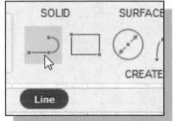

5. In the *Sketch* toolbar, click on the **Line** icon with the left-mouse-button to activate the Line command.

6. Create a closed-region sketch with the starting point aligned to the projected Y-axis as shown below. (Note that the *Pulley* design is symmetrical about a horizontal axis as well as a vertical axis, which allows us to simplify the 2D sketch as shown below.)

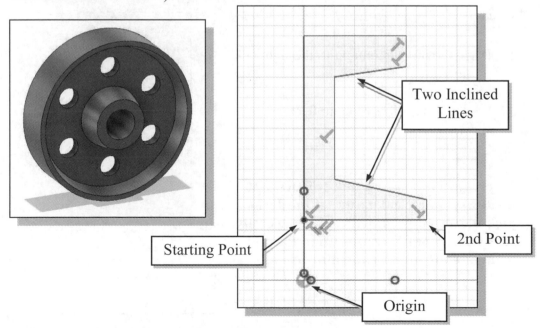

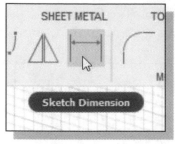

7. Activate the **Sketch Dimension** command in the *Sketch Panel* as shown.

8. Pick the **X-axis** as the first entity to dimension as shown in the figure below.

9. Select the **bottom horizontal line** as the second object to dimension.

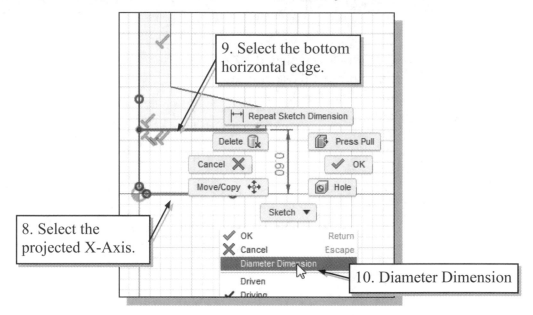

9. Select the bottom horizontal edge.

8. Select the projected X-Axis.

10. Diameter Dimension

10. Inside the graphics window, right-click to bring up the option menu and select **Diameter Dimension**.

11. Place the dimension text to the right side of the sketch.

12. Set the dimension value to **0.75** as shown.

- **To create a dimension that will account for the symmetrical nature of the design, pick the axis of symmetry, pick the entity, select Diameter Dimension in the option menu, and then place the dimension.**

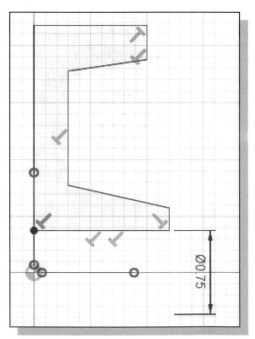

13. Pick the projected **X-axis** as the first entity to dimension as shown in the figure below.

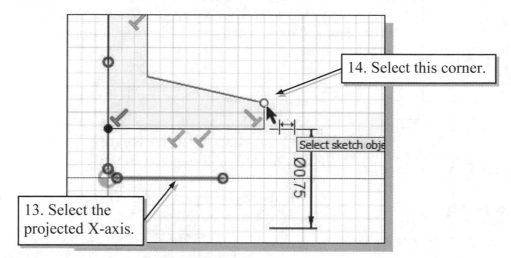

14. Select this corner.

13. Select the projected X-axis.

14. Select the **corner point** as the second object to dimension as shown in the above figure.

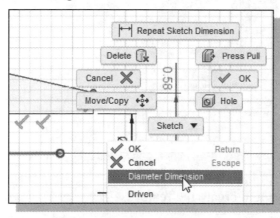

15. Inside the graphics window, right-click to bring up the option menu and select **Diameter Dimension**.

16. Place the dimension text toward the right side of the sketch.

17. Set the dimension value to **1.25**.

18. On your own, create and adjust the vertical size/location dimensions as shown below. (Hint: Create the 4.00 dimension next.)

19. Select **Stop Sketch** in the *Ribbon* toolbar to end the Sketch option.

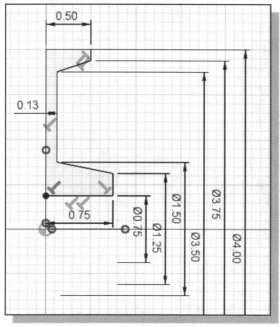

FINISH SKETCH ▾

Create the Revolved Feature

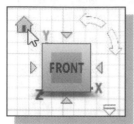

1. Single left-click to activate the **Home View** option as shown. The view will be adjusted back to the default *isometric view.*

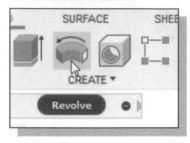

2. In the *Create* panel, select the **Revolve** command by clicking the left-mouse-button on the icon.

3. Confirm the inside region of the 2D sketch is pre-selected as the profile to be revolved.

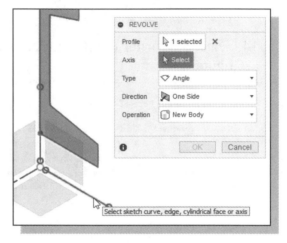

4. In the *Revolve* dialog box, activate the Axis selection and select **X-Axis** as the axis of rotation in the *graphics* window as shown.

5. In the *Revolve* dialog box, confirm the termination *Angle* option is set to **360 deg** as shown. Click **OK** to accept the settings and create the feature.

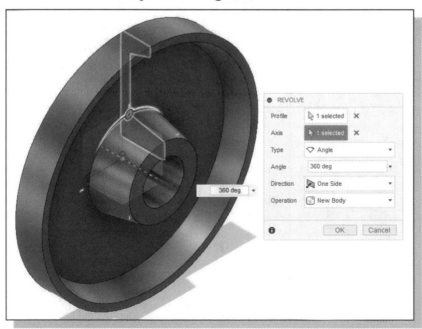

Mirroring Features

In Autodesk Fusion 360, features can be mirrored to create and maintain complex symmetrical features. We can mirror a feature about a work plane or a specified surface. We can create a mirrored feature while maintaining the original parametric definitions, which can be quite useful in creating symmetrical features. For example, we can create one quadrant of a feature, and then mirror it twice to create a solid with four identical quadrants.

1. In the *Create* panel, select the **Mirror** command by clicking the left-mouse-button on the icon.

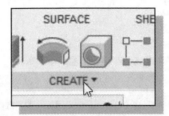 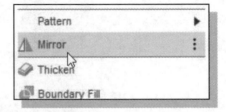

2. In the *Mirror* dialog box, set the *Pattern type* to **Features** and select the 3D base feature in the *Timeline control*.

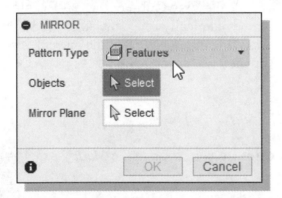 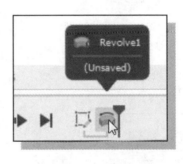

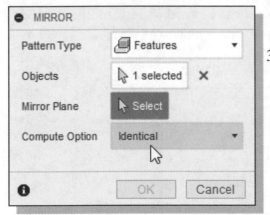

3. In the *Mirror* dialog box, set the Compute Option to **Identical** and click the **Mirror Plane** button to activate the next selection. Autodesk Fusion 360 expects us to select a planar surface about which to mirror.

4. On your own, use the ViewCube or the 3D-Rotate function key to dynamically rotate the solid model so that we are viewing the back surface as shown on the next page.

5. Select the back surface as shown as the planar surface about which to mirror.

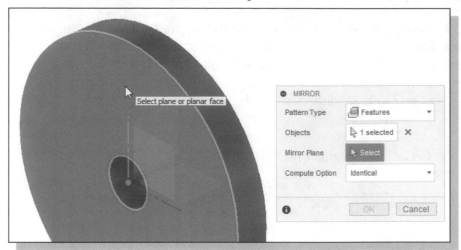

6. Click on the **OK** button to accept the settings and create a mirrored feature.

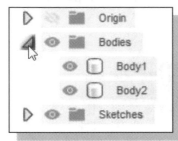

7. On your own, use the ViewCube or the 3D-Rotate function key to dynamically rotate the solid model and view the mirrored feature. Note that the current version of Fusion 360 created two separate bodies.

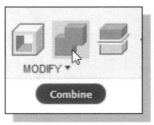

8. Activate the **Combine** command in the *Modify* panel as shown.

    •   The *Combine* command can be used to perform the Boolean operations to join objects into one body.

9. Select the original feature as the *Target body* as shown.

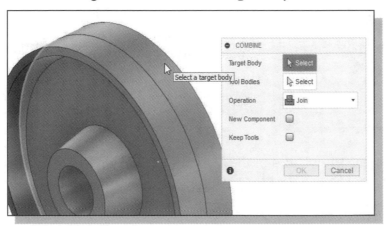

10. Select the copied feature as the *Tool body* as shown.

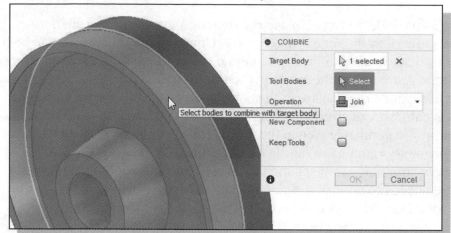

11. In the *Combine* dialog box, confirm the *Operation* is set to **Join** as shown.

12. Click **OK** to accept the settings and create the combined solid model.

13. On your own, use the ViewCube or the 3D-Rotate function key to dynamically rotate the solid model and view the solid model.

> Now is a good time to save the model (quick-key: [**Ctrl**] + [**S**]). It is a good habit to save your model periodically, just in case something might go wrong while you are working on it.

14. Select **Home View** in the option list to reset the display of the model on the screen.

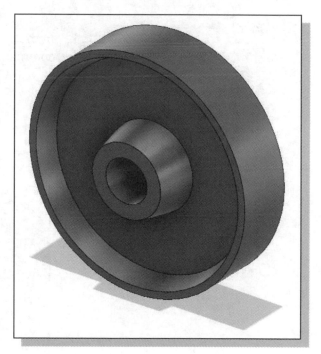

Create a Pattern Leader Using Construction Geometry

In Autodesk Fusion 360, we can also use **construction geometry** to help define, constrain, and dimension the required geometry. **Construction geometry** can be lines, arcs, and circles that are used to line up or define other geometry but are not themselves used as the shape geometry of the model. When profiling the rough sketch, Autodesk Fusion 360 will separate the construction geometry from the other entities and treat them as construction entities. Construction geometry can be dimensioned and constrained just like any other profile geometry. When the profile is turned into a 3D feature, the construction geometry remains in the sketch definition but does not show in the 3D model. Using construction geometry in profiles may mean fewer constraints and dimensions are needed to control the size and shape of geometric sketches. We will illustrate the use of the construction geometry to create a cut feature.

- The *Pulley* design requires the placement of five identical holes on the base solid. Instead of creating the five holes one at a time, we can simplify the creation of these holes by using the **Pattern** command, which allows us to create duplicate features. Prior to using the Pattern command, we will first create a *pattern leader*, which is a regular extruded feature.

1. In the *Sketch* toolbar select the **Start 2D Sketch** command by left-clicking once on the icon.

2. Pick the middle surface of the base feature to align the sketching plane as shown.

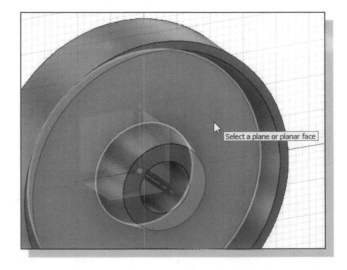

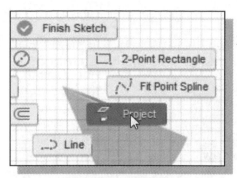

3. Inside the graphics window, **right-click** to bring up the option menu and activate **Sketch → Project** command.

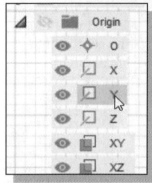

4. Inside the *browser* window, select the **Y-axis** and **Z-axis** to project these entities onto the sketching plane.

5. Click **OK** to accept the selections and create the projected geometry.

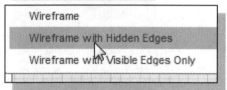

6. On your own, set the display to *wireframe* by clicking on the **Wireframe with Hidden Edges Display** icon in the *View* panel.

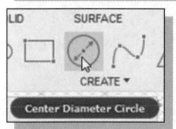

7. Select the **Center Diameter Circle** command by clicking once with the left-mouse-button on the icon in the *Sketch Panel*.

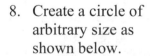

8. Create a circle of arbitrary size as shown below.

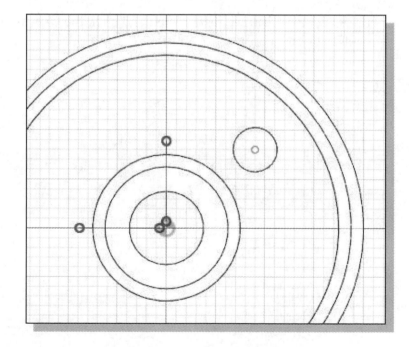

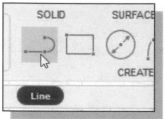

9. In the *Sketch* toolbar, click on the **Line** icon with the left-mouse-button to activate the Line command.

10. Create a *line* by connecting from the center of the circle we just created to the *origin* at the center of the 3D model as shown below.

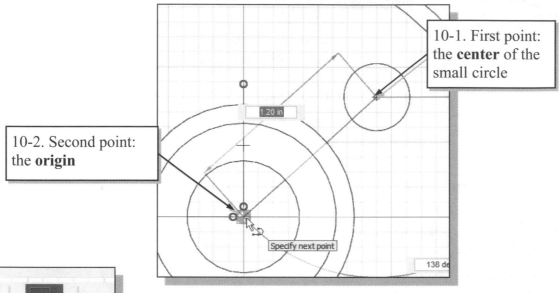

10-1. First point: the **center** of the small circle

10-2. Second point: the **origin**

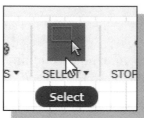

11. Click **Select** in the *Sketch panel* to exit the *Line* command as shown.

12. Select the **Line** we just created and click the **Construction** icon to toggle the line to be a construction entity.

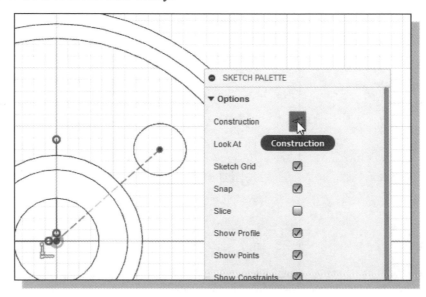

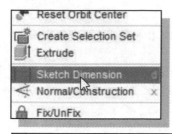

13. Activate the **Sketch Dimension** command in the *option menu* through the right-mouse option as shown.

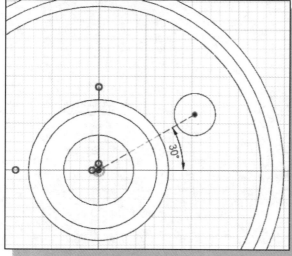

14. Pick the **projected Z axis** as the first entity to dimension as shown in the figure.

15. Select the **construction line** as the second object to dimension.

16. Place the dimension text to the right of the model as shown.

17. On your own, set the angle dimension to **30** as shown in the figure.

➢ Note that the small circle moves as the location of the construction line is adjusted by the *angle dimension* we created.

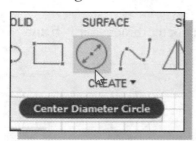

18. Select the **Center Diameter Circle** command by clicking once with the left-mouse-button on the icon in the *Sketch Panel*.

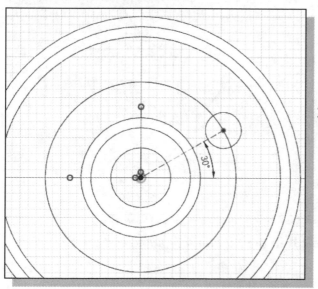

19. Create a *construction circle* by placing the center at the **origin**.

20. Pick the **center** of the small circle to set the size of the construction circle as shown.

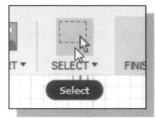

21. Click **Select** in the *Sketch panel* to exit the *Circle* command.

22. Select the Circle we just created and click the **Normal/Construction** icon to toggle the circle to be a construction entity.

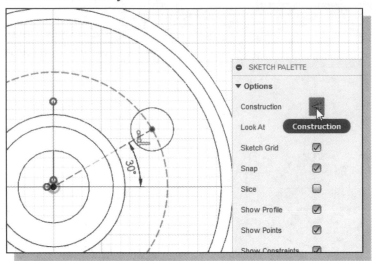

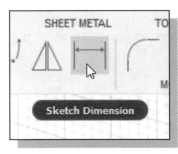

23. Activate the **Sketch Dimension** command in the *Sketch Panel* as shown.

24. On your own, create the two diameter dimensions, **.5** and **2.5**, as shown below.

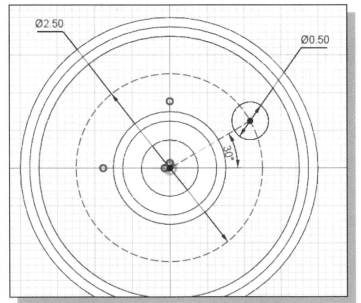

25. Select **Stop Sketch** in the *Ribbon toolbar* to end the Sketch option.

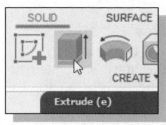

26. In the *Create Features* panel select the **Extrude** command by left-clicking on the icon.

27. Pick the inside region of the circle to set up the profile of the extrusion.

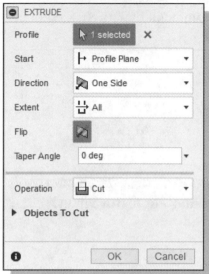

28. Drag the arrow so that the cut direction is into the solid model.

29. Inside the *Extrude* dialog box, select the **Cut** operation and set the *Extent* to **All** as shown.

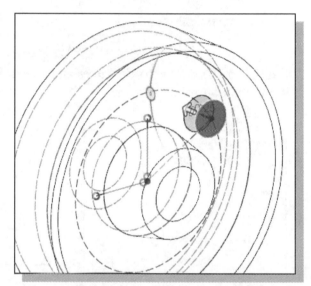

30. Click on the **OK** button to accept the settings and create the cut feature.

31. On your own, adjust the angle dimension applied to the construction line of the cut feature to **90** and observe the effect of the adjustment.

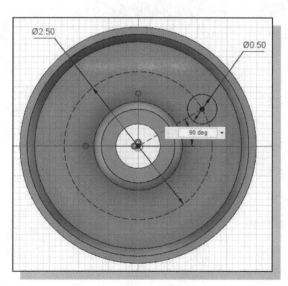

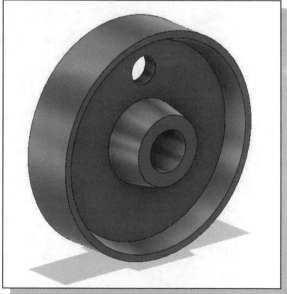

Circular Pattern

In Autodesk Fusion 360, existing features can be easily duplicated. The **Pattern** command allows us to create both rectangular and polar arrays of features. The patterned features are parametrically linked to the original feature; any modifications to the original feature are also reflected in the arrayed features.

1. In the *Create* panel, select the **Circular Pattern** command by left-clicking once on the icon.

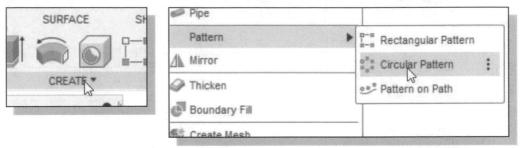

2. Set the *Pattern Type* to **Features** and select the **circular cut feature** in the Timeline control as shown.

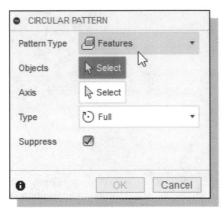

3. In the *Circular Pattern* dialog box, set the Compute Option to **Identical**.

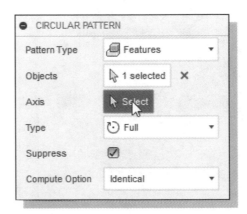

4. In the *Circular Pattern* dialog box, click the **Axis** button.

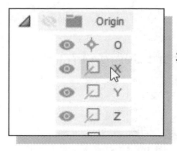

5. Select the **X-Axis** in the *browser* window or in the graphics window.

6. In the *Circular Pattern* dialog box, set the *Type* to **Full** and enter **6** in the *quantity* box as shown.

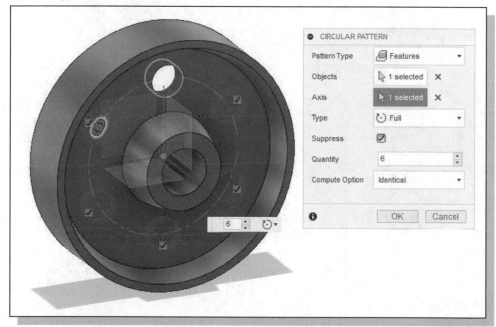

7. Click on the **OK** button to accept the settings and create the *circular pattern*.

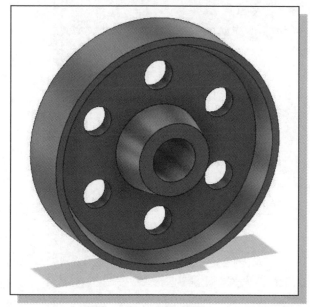

8. Select **Save** in the *Standard* toolbar; we can also use the "**Ctrl-S**" combination (press down the [Ctrl] key and hit the [S] key once) to save the part as ***Pulley*** in the **Chapter11** folder.

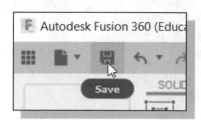

Start a New 2D Drawing

1. In the *File* pull down menu, select **New Drawing → From Design** to start a new drawing.

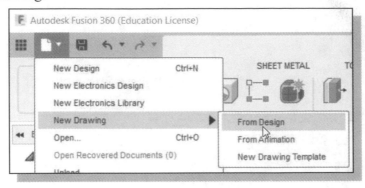

2. In the **Create Drawing** dialog box, set the Standard to ASME, units to in and sheet size to B (17in x 11 in) as shown and click OK.

➤   Note that a new tab appears in the Fusion 360 window. We can switch between the solid model and the drawing by clicking the corresponding tabs in the ribbon toolbar area.

❖   In the graphics window, Autodesk Fusion 360 displays a default drawing sheet that includes a title block.

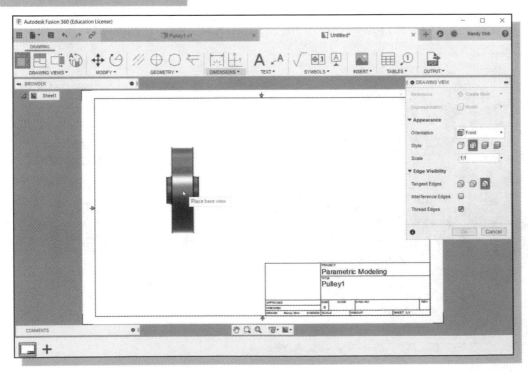

Add a Base View

In Autodesk Fusion 360 *Drawing Mode*, the first drawing view we create is called a **base view**. A *base view* is the primary view in the drawing; other views can be derived from this view. When creating a *base view*, Autodesk Fusion 360 allows us to specify the view to be shown. By default, Autodesk Fusion 360 will treat the *world XY plane* as the front view of the solid model. Note that there can be more than one *base view* in a drawing.

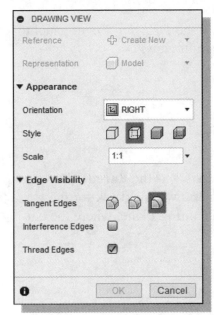

1. In the *Drawing View* dialog box, set *Orientation* to **Right** view, set the **Scale** to **1 : 1** and *Style* to **Hidden Line** as shown in the figure.

2. Confirm the *Tangent Edges* option is set to **off** as shown.

3. Inside the graphics window, place the **base view** toward the left side of the drawing sheet as shown below. (If necessary, drag the *Drawing View* dialog box to another location on the screen.)

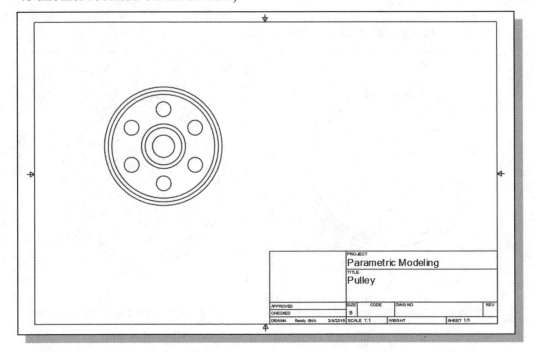

4. Click the [**OK**] key once to exit the *Drawing View* dialog box.

Create a Section View

Section views are used to make a part drawing more understandable, showing the internal details of the part. Since the sectioned drawing is showing the internal features there is generally no need to show hidden lines. In Autodesk Fusion 360 *Drawing Mode*, the **Section View** command is used to create *Section* views.

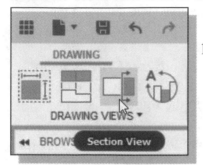

1. Click on **Section View** in the *Drawing Views* panel to create a new projected section view.

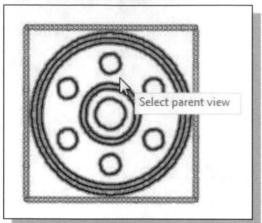

2. Select the *base view* as the **Parent View** as shown. Fusion 360 will next ask for the definition of a Cutting Plane, where the cut shall happen.

3. Move the cursor on top of the **center** of the base view to activate alignment and create a ***cutting plane line*** as shown.

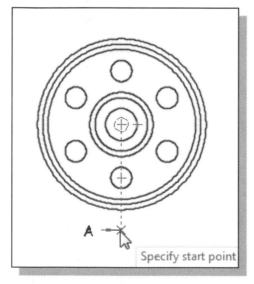 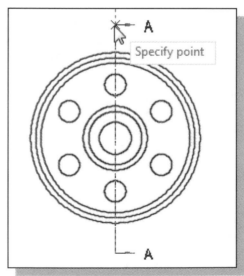

4. Press the [**Enter**] key once to accept the placement of the cutting plane line.

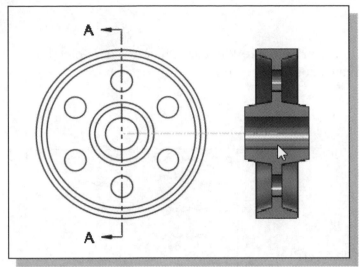

5. Move the cursor toward the **right side** of the base view and place the section view as shown.

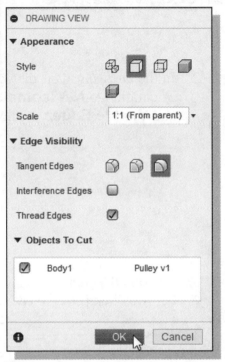

6. In the *Drawing View* dialog box, set *Style* to **Visible Edges**, set the **Scale** to **1 : 1 (From Parent)** and **Tangent Edges** to **off** as shown in the figure.

7. Click **OK** to create the section view.

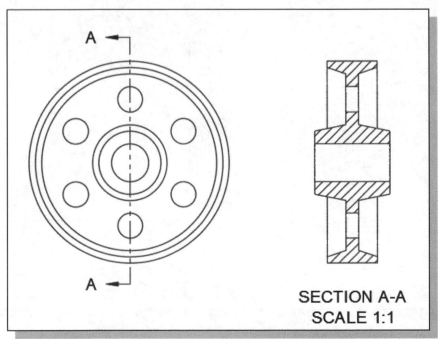

SECTION A-A
SCALE 1:1

Create an Isometric View

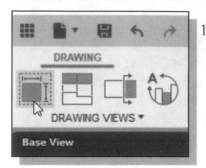

1. Click on **Base View** in the *Drawing Views* panel to create another base view.

2. In the *Drawing View* dialog box, set orientation to **NW Isometric**, **Scale 1 : 1** and **Visible Edges** as shown in the figure.

3. Position the *isometric* view toward the right side of the drawing sheet as shown below. Click [**OK**] to create the view.

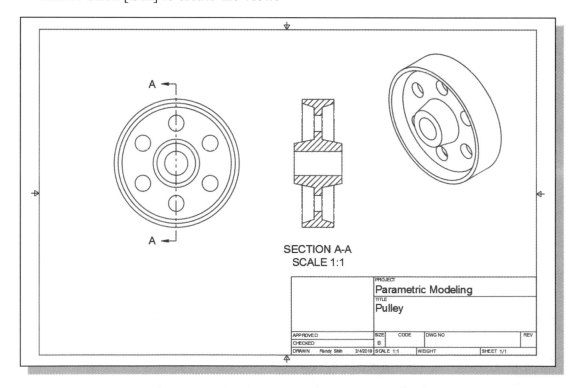

Add Center Mark on Patterned feature

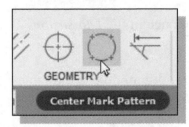

1. Click on the **Center Mark Pattern** button in the *Centerlines* panel as shown.

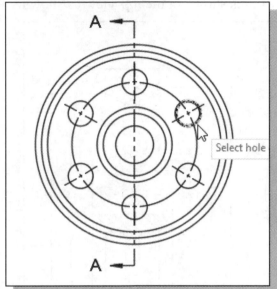

2. Select the six patterned circles in the base view to create the center mark pattern as shown.

3. In the **Center Mark Pattern** dialog box, switch on the **Full PCD** and **Center Mark** options as shown.

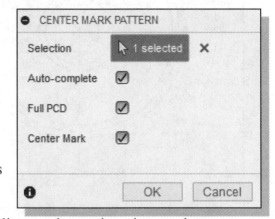

4. Click [**OK**] to create the center marks as shown.

5. On your own, create additional centerlines to the section view as shown.

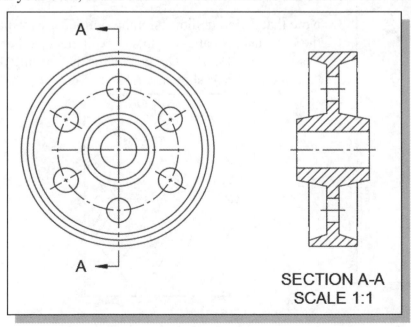

SECTION A-A
SCALE 1:1

Complete the Drawing Sheet

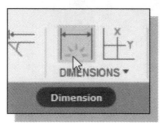

1. Select **Dimension** by left-clicking once in the *Dimensions* toolbar panel.

   • The **Dimension** command in the drafting module is the same as the dimension command in the modeling module.

2. Create the dimensions for the different features shown in the *top* view as shown.

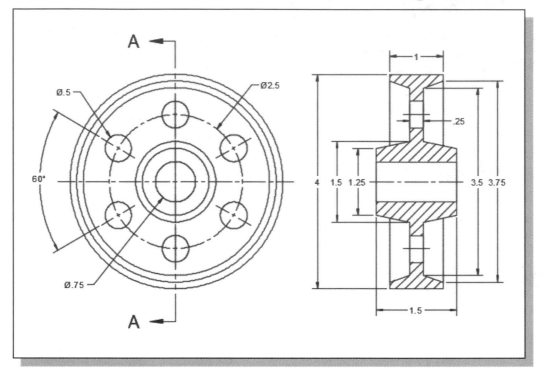

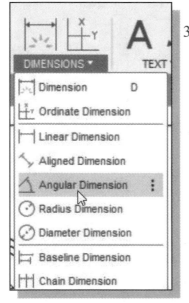

3. Note that different dimensioning options are available in the Dimensions panel for specific dimensions. For example, the *Angular Dimension* for the spacing of the patterned holes as shown.

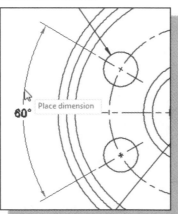

4. In the *Text Panel*, click on the **Text** button.

5. Pick a location that is inside the *DWG No* area and specify a text box region as shown.

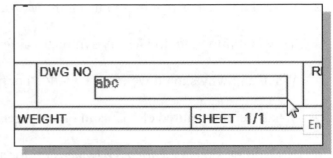

6. Enter **MECH 375-CH11-01** as the new *Drawing No* as shown.

7. On your own, complete the multi-view drawing and print out a copy of the drawing as shown.

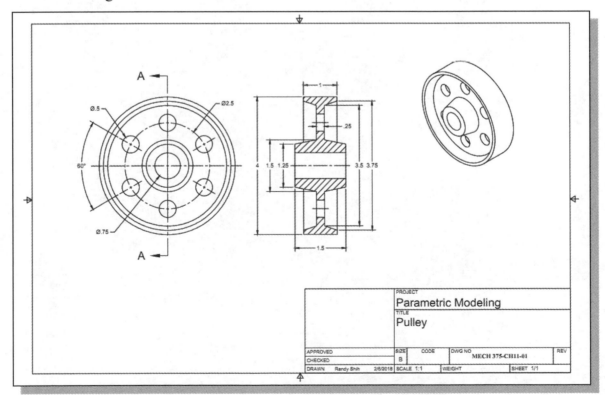

Review Questions:

1. List the different symmetrical features created in the *Pulley* design.

2. Describe the steps required in using the **Mirror Feature** command.

3. Why is it important to identify symmetrical features in designs?

4. When and why should we use the **Pattern** option?

5. What are the required elements in order to generate a sectional view?

6. How do we create a *Diameter Dimension* for a revolved feature?

7. What is the difference between *construction geometry* and *normal geometry*?

8. What is the main difference between a base view and a projected view?

Exercises: Create the Solid models and save the exercises in the Chapter 11 folder.

1. **Shaft Support** (Dimensions are in inches.)

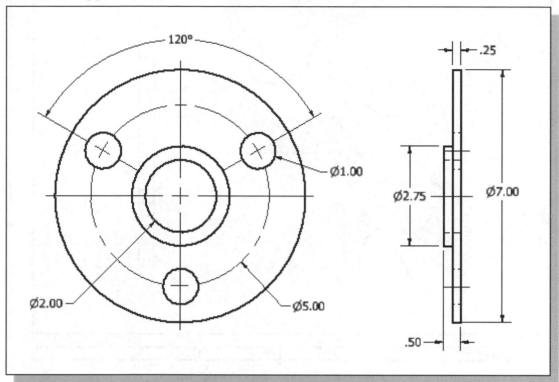

2. **Ratchet Plate** (Dimensions are in inches. Thickness: 0.125 inch.)

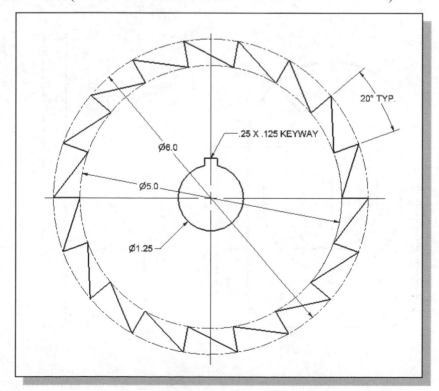

3. **Geneva Wheel** (Dimensions are in inches.)

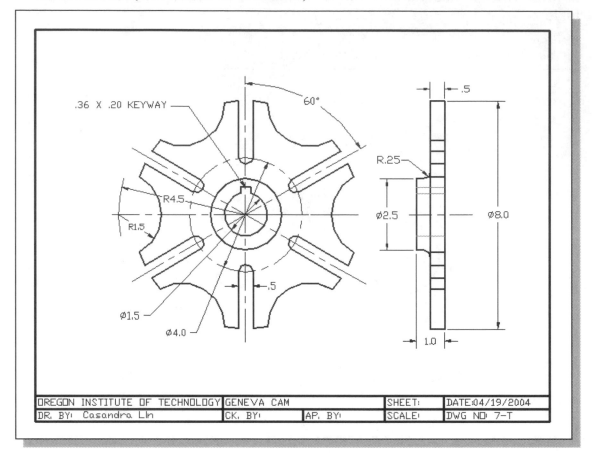

OREGON INSTITUTE OF TECHNOLOGY	GENEVA CAM		SHEET:	DATE:04/19/2004
DR. BY: Casandra Lin	CK. BY:	AP. BY:	SCALE:	DWG NO: 7-T

4. **Support Mount** (Dimensions are in inches.)

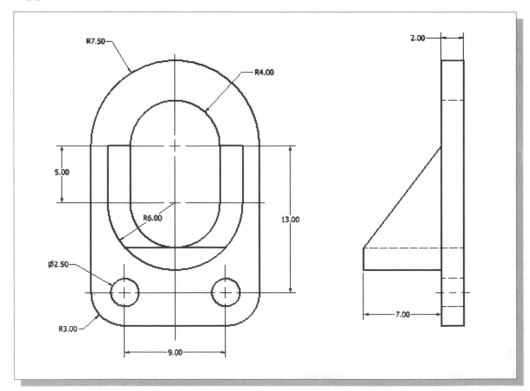

5. **Hub** (Dimensions are in inches.)

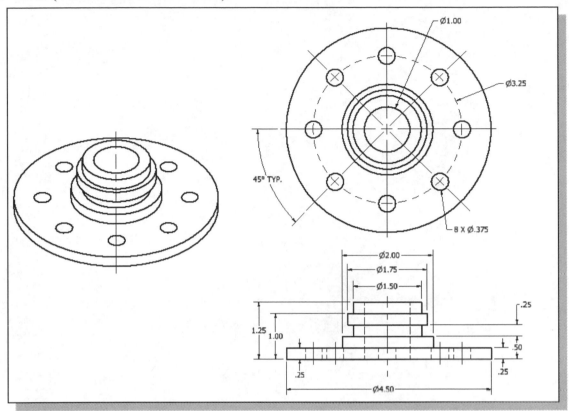

6. **Switch Base** (Dimensions are in inches.)

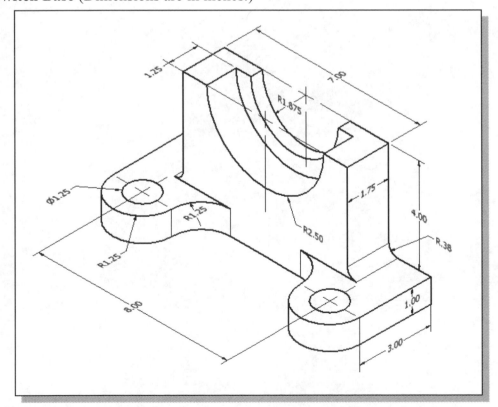

Notes:

Chapter 12
Advanced 3D Construction Tools

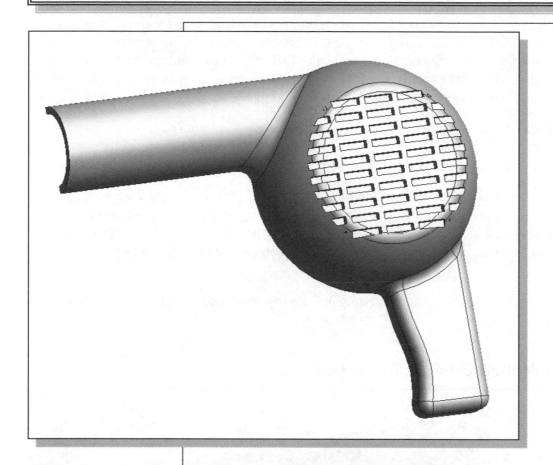

Learning Objectives

- **Understand the Concepts Behind the Different 3D Construction Tools**
- **Set up Multiple Work Planes**
- **Create Swept Features**
- **Create Lofted Features**
- **Use the Shell Command**
- **Create 3D Rounds & Fillets**

Introduction

Autodesk Fusion 360 provides an assortment of three-dimensional construction tools to make the creation of solid models easier and more efficient. As demonstrated in the previous lessons, creating **extruded** features and **revolved** features are the two most common methods used to create 3D models. In this next example, we will examine the procedures for using the **Sweep** command, the **Loft** command, and the **Shell** command, and also for creating **3D rounds** and **fillets** along the edges of a solid model. These types of features are common characteristics of molded parts.

The **Sweep** option is defined as moving a cross-section through a path in space to form a three-dimensional object. To define a sweep in Autodesk Fusion 360, we define two sections: the trajectory and the cross-section.

The **Loft** command allows us to blend multiple profiles with varying shapes on separate planes to create complex shapes. Profiles are usually on parallel planes; however, non-parallel planes can also be used. We can use as many profiles as we wish but, to avoid twisting the loft shape, we should map points on each profile that align along a straight vector.

The **Shell** option is defined as hollowing out the inside of a solid, leaving a shell of specified wall thickness.

A Thin-Walled Design: Dryer Housing

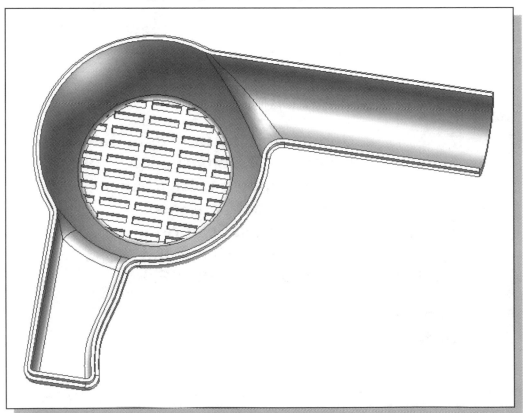

Modeling Strategy

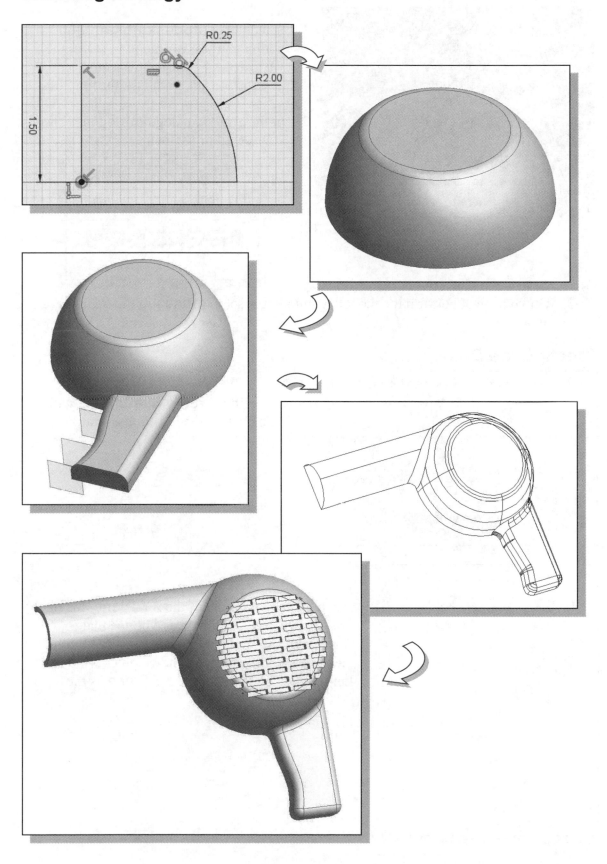

Starting Autodesk Fusion 360

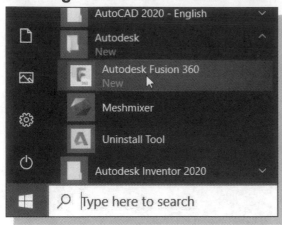

1. Select the **Autodesk Fusion 360** option on the *Start* menu or select the **Autodesk Fusion 360** icon on the desktop to start Autodesk Fusion 360.

2. In the *Sign In* dialog box, log in with your email or username.

3. On your own, confirm the modeling units set is set to **Inches** in the *Browser* area.

Creating the Base Feature

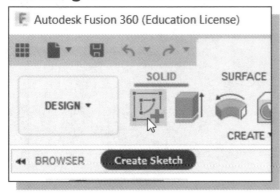

1. Activate the **Create Sketch** icon with a single click of the left-mouse-button.

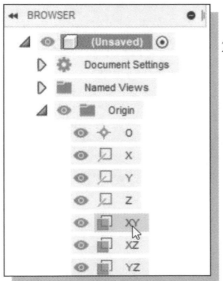

2. Move the cursor on top of the **XY Plane**, inside the *browser* window as shown, and notice that Autodesk Fusion 360 will automatically highlight the corresponding plane in the graphics window. Left-click once to select the XY Plane as the sketching plane.

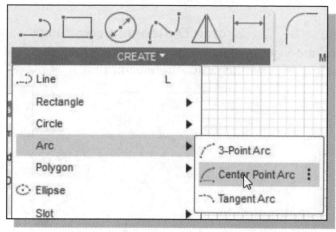

3. Activate the **Center Point Arc** command by clicking once with the left-mouse-button on the icon in the *Sketch Panel*.

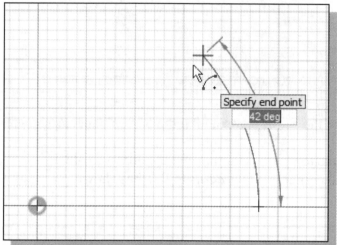

4. Pick the **Origin** as the center location of the new circle.

5. Move the cursor toward the right and click on the x axis to align the first end point of the arc to the origin horizontally.

6. Create an **arc** of arbitrary size as shown.

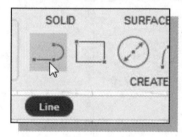

7. Click on the **Line** icon in the *2D Sketch* panel.

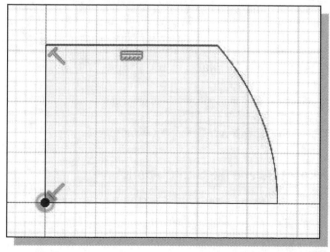

8. Create the three line segments as shown. The lines are either horizontal or vertical; the lines are also attached to the arc with the **lower left corner** aligned to the **Origin**.

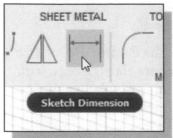

9. On your own, use the **Sketch Dimension** command to create and modify the dimensions as shown in the figure below. (Hint: Modify the radius dimension first.)

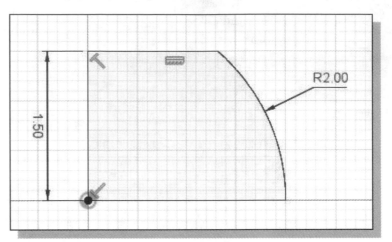

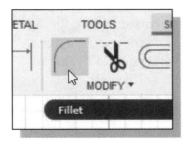

10. Activate the **Fillet** command in the *Sketch* panel as shown.

11. Select the top horizontal line and the arc to create a rounded corner; set the *radius* to **0.25** as shown.

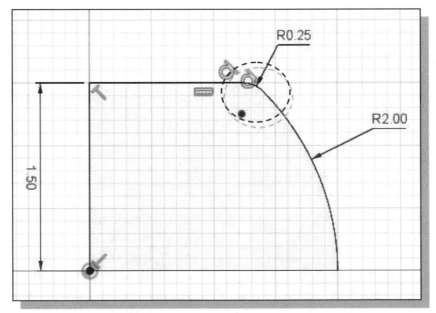

12. Select **Finish Sketch** to end the Sketch option.

Create a Revolved Feature

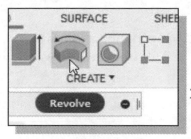

1. In the *Create* panel, select the **Revolve** command by clicking the left-mouse-button on the icon.

2. Confirm the inside region of the 2D sketch is pre-selected as the profile to be revolved.

3. In the *Revolve* dialog box, activate the **Axis selection** and select **Y-Axis** as the axis of rotation in the *graphics* window as shown.

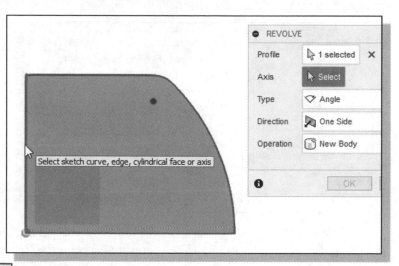

4. In the *Revolve* dialog box, set the termination *Extents* option to **360** as shown.

5. Click on the **OK** button to accept the settings and create the revolved feature.

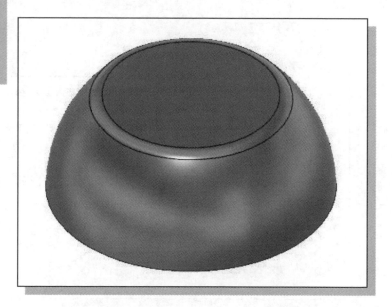

Create Offset Work Planes

1. Single left-click to activate the **Home View** option as shown. The view will be adjusted back to the default *isometric view*.

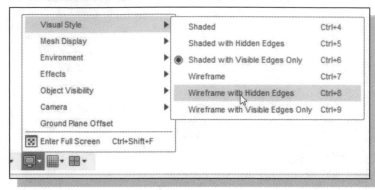

2. Select **Wireframe with Hidden edges** display mode under the *Visual Style* option as shown.

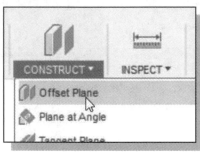

3. In the *Construct* panel, select the **Offset Plane** command by left-clicking the icon.

    •   Autodesk Fusion 360 expects us to select any existing geometry, which will be used as a reference to create the new work plane.

4. Select the **XY Plane** in the *browser* window as the reference plane. Left-click once to set the XY Plane as the reference of a new work plane.

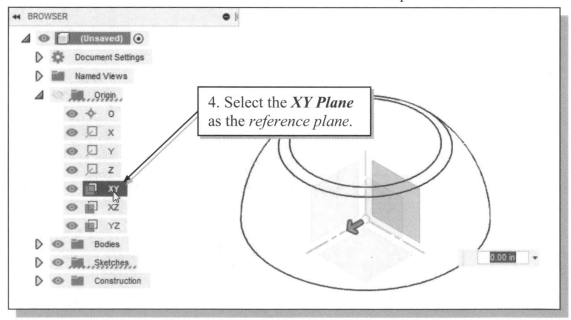

4. Select the *XY Plane* as the *reference plane*.

5. In the *Offset Plane* dialog box, enter **2.5** as the offset distance for the new work plane.

6. Click on the **OK** button to accept the setting and create a work plane.

7. On your own, repeat the above steps and create two additional work planes that are **3.5** inches and **4.25** inches away from the XY Plane.

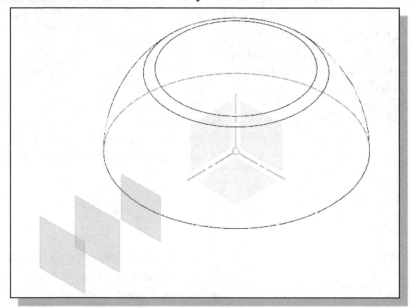

Start 2D Sketches on the Work Planes

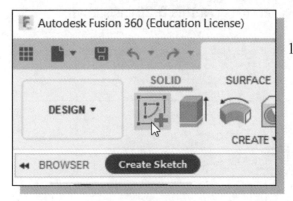

1. Activate the **Create Sketch** icon with a single click of the left-mouse-button.

2. Select the **XY Plane** in the *browser* window.

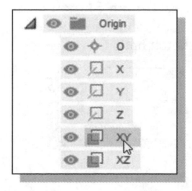

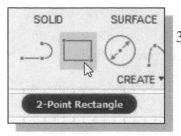

3. Select the **2-point rectangle** command by clicking once with the left-mouse-button on the icon in the *Sketch* toolbar.

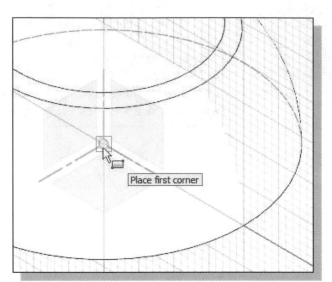

4. Click on the **Origin** to align the first corner of the rectangle.

5. Create a rectangle of arbitrary size by selecting a location that is toward the right side of the graphics window as shown.

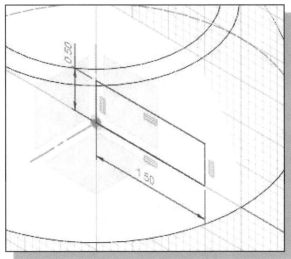

6. On your own, create and modify the two dimensions, **0.5x1.5**, as shown in the figure.

7. On your own, create two rounded corners (**radius 0.25**) as shown.

8. Select **Finish Sketch** in the *Ribbon* toolbar to end the Sketch option.

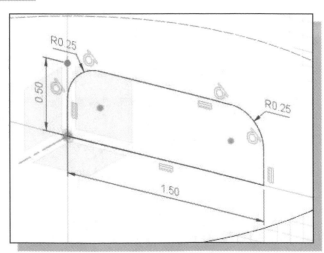

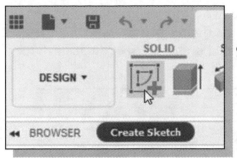

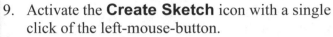

9. Activate the **Create Sketch** icon with a single click of the left-mouse-button.

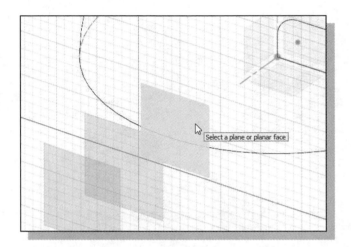

10. Select **WorkPlane1**, by left-clicking once on the first offset work plane in the graphics window as shown.

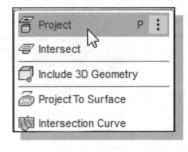

11. Select the **Project Geometry** command in the *Sketch* panel.

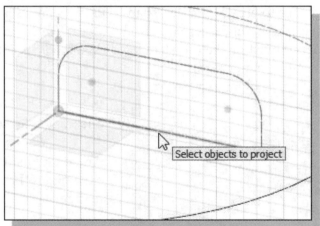

12. Select the bottom edge of the 2D sketch we just created, as shown in the figure.

13. Click **OK** to accept the selection.

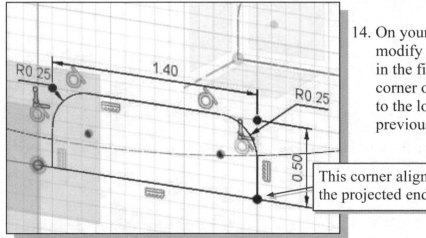

14. On your own, create and modify the 2D sketch as shown in the figure. (The lower right corner of the sketch is aligned to the lower right corner of the previous sketch.)

This corner aligned to the projected endpoint.

15. On your own, repeat the above steps and create two additional sketches on **WorkPlane2** and **WorkPlane3** as shown in the figures below. Note that the lower right corners of the four sketches are aligned in the Z-direction through the use of the projected bottom edge.

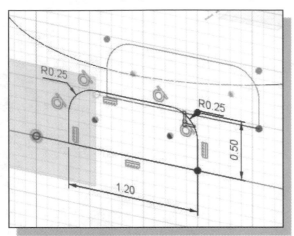
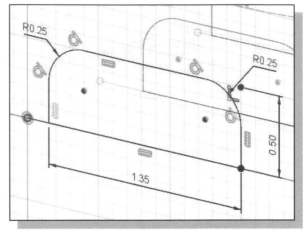

Create a Lofted Feature

The **Loft** option allows us to blend multiple profiles with varying shapes on separate planes to create complex shapes. Profiles are usually on parallel planes, but any non-perpendicular planes can also be used. We can use as many profiles as we wish, but to avoid twisting the loft shape, we should map points on each profile that align along a straight vector.

1. In the *Create Panel*, select the **Loft** command by left-clicking on the icon.

• Autodesk Fusion 360 expects us to select a number of existing profiles, which will be used to create the lofted feature.

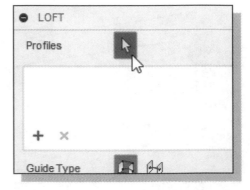

• In the *Loft* dialog box, the **Profiles** selection box is used to manage the selected multiple sketches.

2. Pick the four sketched sections, in the order that they were created, by selecting the **inside** of the sketches in the Graphics window.

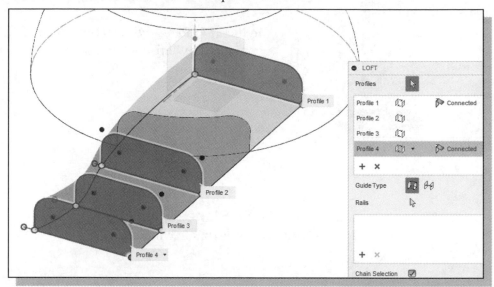

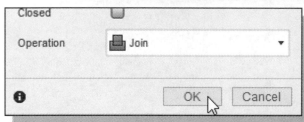

3. Set the Operation option to **Join** and click the **OK** button to accept the settings and create the lofted feature.

Create an Extruded Feature

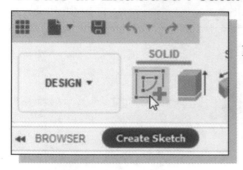

1. In the *3D Model tab* select the **Start 2D Sketch** command by left-clicking once on the icon.

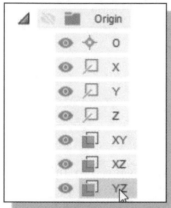

2. Select the **YZ Plane**, by left-clicking once on the YZ Plane, in the *browser* window.

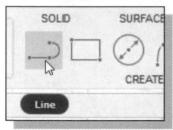

3. Click on the **Line** icon in the *2D Sketch* panel.

4. Create a line that is parallel (nearly aligned) to the horizontal axis as shown.

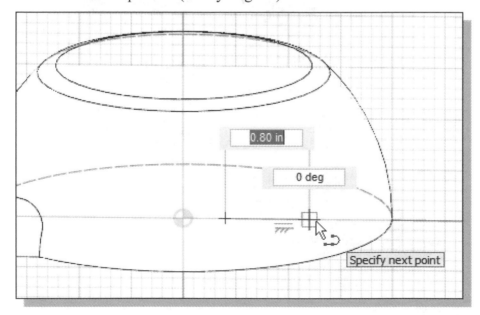

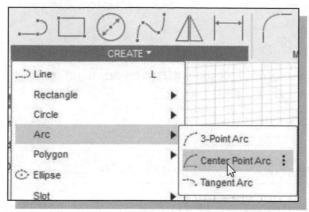

5. Select the **Center point arc** option in the *2D Sketch* panel as shown.

6. On your own create an arc center aligned to the **mid-point** and **endpoints** of the previously created line as shown in the figure below.

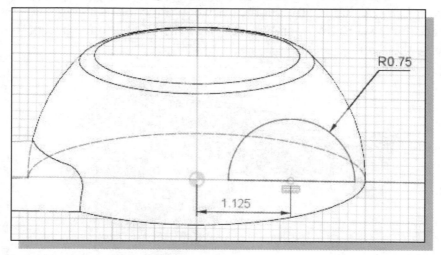

7. On your own, create and modify the two dimensions, and apply a coincident constraint as shown in the figure below.

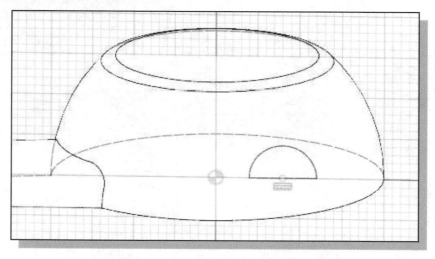

8. Select **Finish Sketch** in the *Ribbon* toolbar to end the Sketch option.

Complete the Extruded Feature

1. In the *3D Model tab*, select the **Extrude** command by left-clicking once on the icon.

2. In the *Extrude* pop-up window, enter -**5.5** as the extrusion distance.

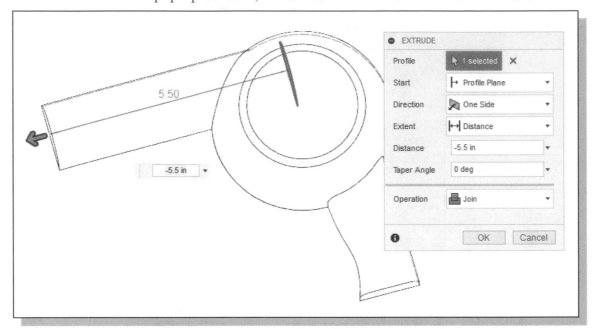

3. In the *Extrude* dialog box, set the extrusion direction as shown and confirm the operation is set to **Join**, and then click on the **OK** button to accept the settings to create the feature.

Create 3D Rounds and Fillets

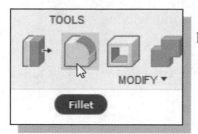

1. In the *Modify* toolbar, select the **Fillet** command by left-clicking once on the icon.

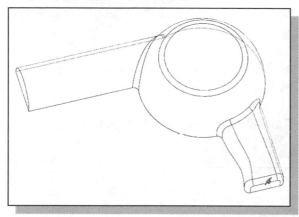

2. Click on the **three edges**, at the intersections of the different features, as shown in the figure. (Hold down the [**Ctrl**] key to select multiple edges.)

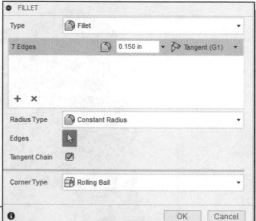

3. In the *Fillet* dialog box, set the *Radius* to **0.15** as shown.

4. Click on the **OK** button to accept the settings and create the 3D rounds and fillets.

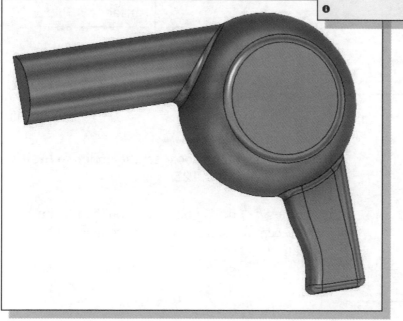

Create a Shell Feature

The **Shell** command can be used to hollow out the inside of a solid, leaving a shell of specified wall thickness.

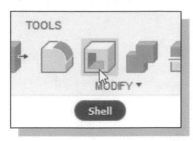

1. In the *Modify* toolbar, select the **Shell** command by left-clicking once on the icon.

2. On your own, use the **ViewCube** or the **3D Rotation** quick-key to display the back faces of the model as shown below.

3. In the *Shell* dialog box, the **Remove Faces** option is activated. Select the two faces as shown below.

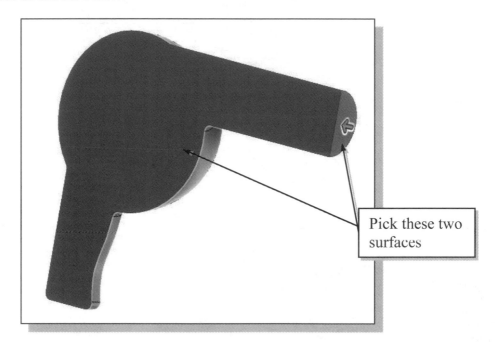

Pick these two surfaces

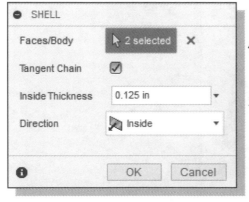

4. In the *Shell* dialog box, set the option to **Inside** with a value of **0.125** as shown.

5. In the *Shell* dialog box, click on the **OK** button to accept the settings and create the shell feature.

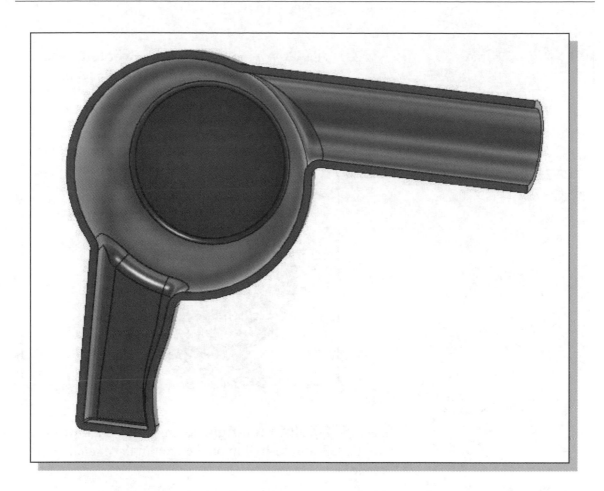

Create a Pattern Leader

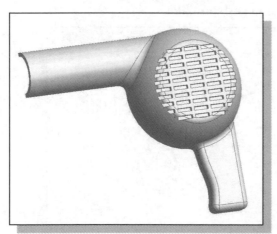

- The *Dryer Housing* design requires the placement of identical holes on the top face of the solid. Instead of creating the holes one at a time, we can simplify the creation of these holes by using the **Pattern** command to create duplicate features. Prior to using the Pattern command, we will first create a *pattern leader*, which is a regular cut feature.

1. Select **Home View** in the **ViewCube** to adjust the display of the model to the isometric view angle.

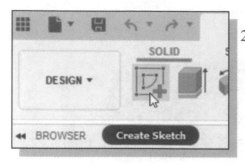

2. In the *Sketch* toolbar select the **Create Sketch** command by left-clicking once on the icon.

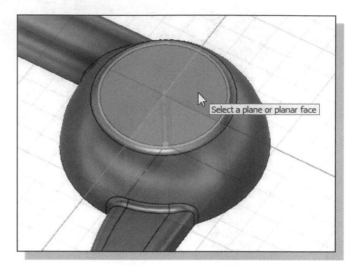

3. Pick the top face of the base feature as shown.

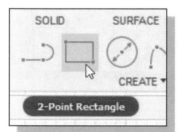

4. Select the **2-point rectangle** command by clicking once with the left-mouse-button on the icon in the *Sketch* toolbar.

5. Create a rectangle and modify the dimensions as shown.

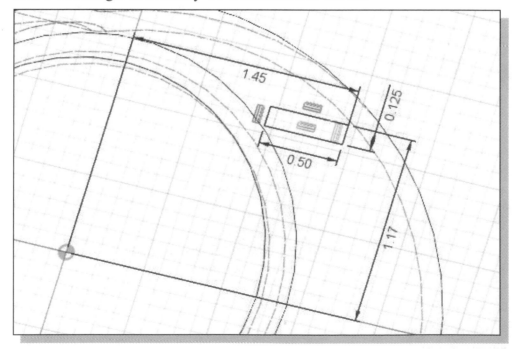

6. Select **Finish Sketch** in the *Ribbon* toolbar to end the Sketch option.

7. In the *3D Model tab*, select the **Extrude** command by left-clicking on the icon.

8. Confirm the inside region of the rectangle is pre-selected as the profile of the extrusion.

9. Inside the *Extrude* dialog box, select the **Cut** operation and set the *Extents* **Distance** to **0.13 in** as shown.

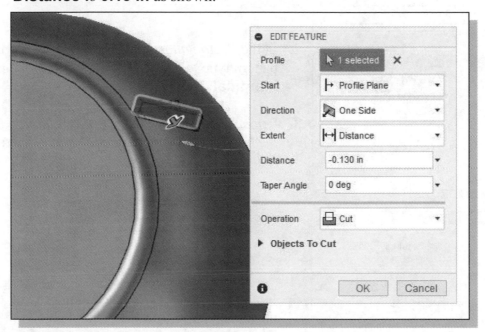

10. In the *Extrude* dialog box, click on the **OK** button to proceed with creating the cut feature.

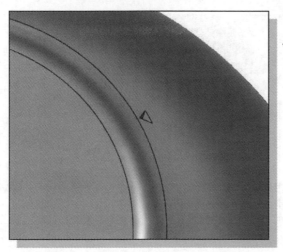

♦ Note that the pattern leader is a fairly small cut feature on the solid model.

Create a Rectangular Pattern

In Autodesk Fusion 360, existing features can be easily duplicated. The **Pattern** command allows us to create both rectangular and polar arrays of features. The patterned features are parametrically linked to the original feature; any modifications to the original feature are also reflected on the arrayed features.

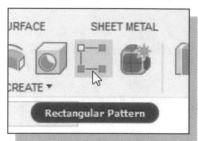

1. In the *Create* toolbar, select the **Rectangular Pattern** command by left-clicking once on the icon.

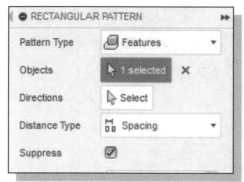

2. In the *Rectangular Pattern* dialog box, set the pattern type to **Features** and *Compute option* to **Identical** as shown.

3. Select **Extrusion2**, the *cut feature created in the last section,* in the *Timeline control* area.

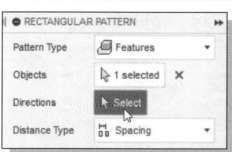

4. In the *Rectangular Pattern* dialog box, set the distance type to **Spacing** and click on the **Directions** icon as shown.

5. Select the two straight edges as shown.

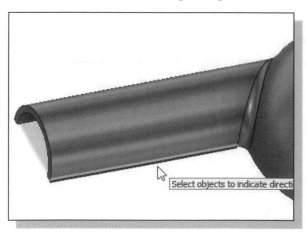

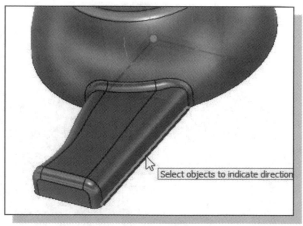

6. In the *Rectangular Pattern* dialog box, enter **0.6** in the *Distance* box and **5** in the *Count* box as shown.

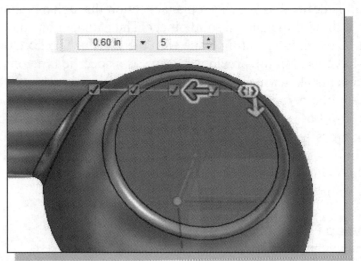

7. In the *Rectangular Pattern* dialog box, set the *Distance type* to **Spacing** as shown.

8. Enter **10** in the *Count* box and **0.25** in the *Spacing* box as shown.

9. Click on the **OK** button to accept the settings and create the *rectangular pattern*.

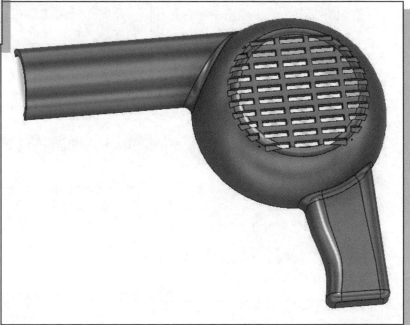

Create a Swept Feature

The **Sweep** operation is defined as moving a planar section through a planar (2D) or 3D path in space to form a three-dimensional solid object. The path can be an open curve or a closed loop but must be on an intersecting plane with the profile. The **Extrusion** operation, which we have used in the previous lessons, is a specific type of sweep. The **Extrusion** operation is also known as a *linear sweep* operation, in which the sweep control path is always a line perpendicular to the two-dimensional section. Linear sweeps of unchanging shape result in what are generally called *prismatic solids* which means solids with a constant cross-section from end to end. In Autodesk Fusion 360, we create a *swept feature* by defining a path and then a 2D sketch of a cross section. The sketched profile is then swept along the planar path. The **Sweep** operation is used for objects that have uniform shapes along a trajectory.

♦ **Define a Sweep Path**

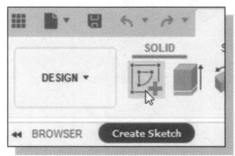

1. In the *3D Model tab* select the **Start 2D Sketch** command by left-clicking once on the icon.

2. Select the **bottom flat face of the model**, by left-clicking once, as shown.

3. Inside the graphics window, right-click to bring up the option menu and select **Project Geometry**.

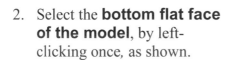

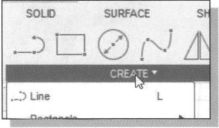

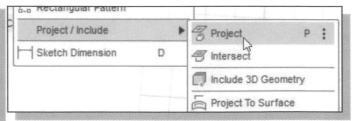

4. Select the **outer edges** of the bottom surface of the model as shown. (Hint: Use the Dynamic Zoom function to assist selecting all neighboring edges.)

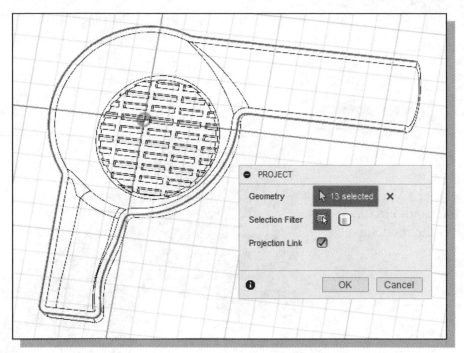

5. Inside the *Project* dialog box, select **OK** to end the Project Geometry command.

6. Select **Finish Sketch** in the *Ribbon toolbar* to end the Sketch option.

*   The projected geometry will be used as the 2D sweep path of the feature.

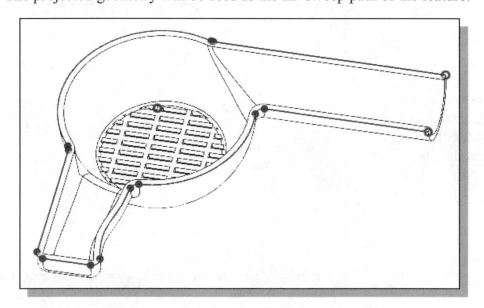

♦ Define the Sweep Section

1. In the *3D Model tab* select the **Start 2D Sketch** command by left-clicking once on the icon.

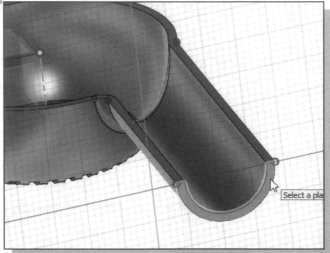

2. Select the small circular surface of the model as shown.

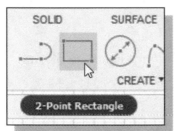

3. Select the **Two point rectangle** command by clicking once with the left-mouse-button on the icon in the *2D Sketch* panel.

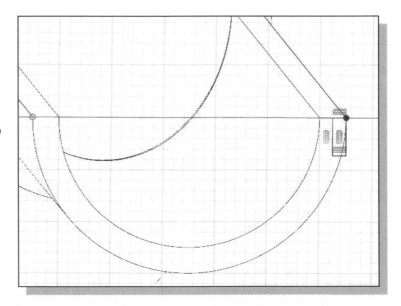

4. On your own, create a rectangle that is aligned to the outside corner of the surface as shown.

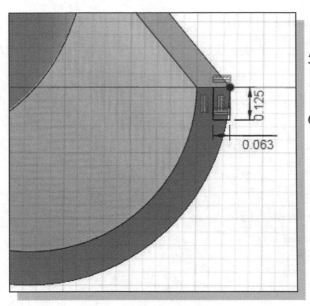

5. On your own, create and modify the dimensions as shown in the figure.

6. Select **Finish Sketch** in the *Ribbon* toolbar to end the Sketch option.

- The created 2D sketch will be used as the sweep section of the feature.

♦ **Complete the Swept Feature**

1. In the *Create toolbar*, select the **Sweep** command by left-clicking once on the icon.

2. Select the region defined by the rectangle we just created as the profile of the sweep operation. (Be sure to select the entire rectangular region.)

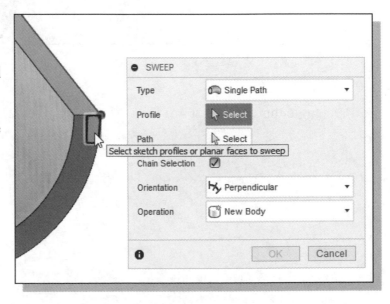

3. In the *Sweep* dialog box, set the *Operation* to **Cut** and activate the **Path** selection option as shown.

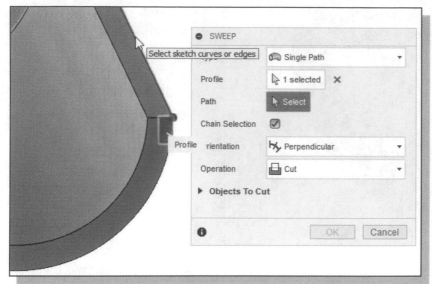

4. Click on the projected open curve we created as the sweep path as shown in the above figure. (Make sure to select the sketch, not the surface.)

5. In the *Sweep* dialog box, click on the **OK** button to accept the settings and create the *swept* feature.

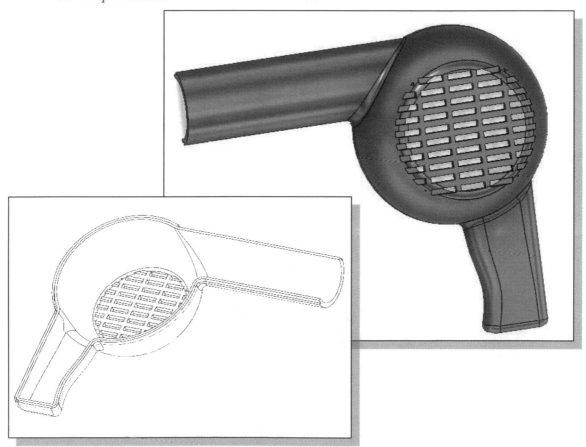

Review Questions:

1. Keeping the *History Tree* in mind, what is the difference between *cut with a pattern* and *cut each one individually*?

2. What is the difference between **Sweep** and **Extrude**?

3. What are the advantages and disadvantages of creating fillets using the **3D Fillets** command and creating fillets in the 2D profiles?

4. Describe the steps used to create the *Shell* feature in the lesson.

5. How do we modify the *Pattern* parameters after the model is built?

6. Describe the elements required in creating a *Swept* feature.

7. Create sketches showing the steps you plan to use to create the model shown on the next page:

Exercises: Create and save the exercises in the Chapter12 folder.

1. **Motor Housing** (Dimensions are in inches.)

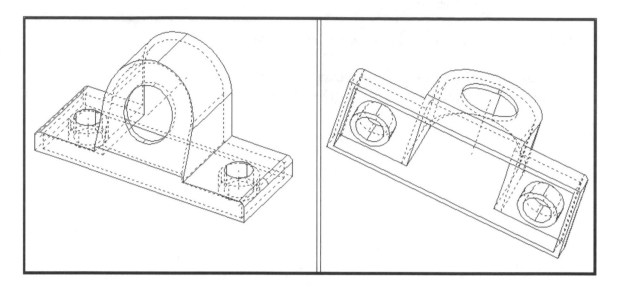

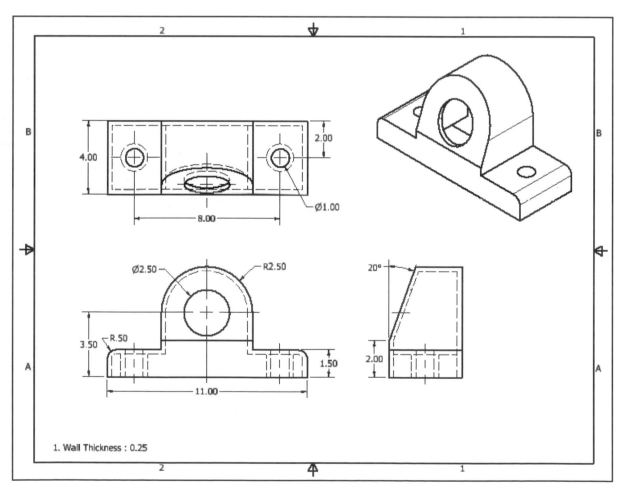

1. Wall Thickness : 0.25

2. **Guide Base** (Dimensions are in mm.)

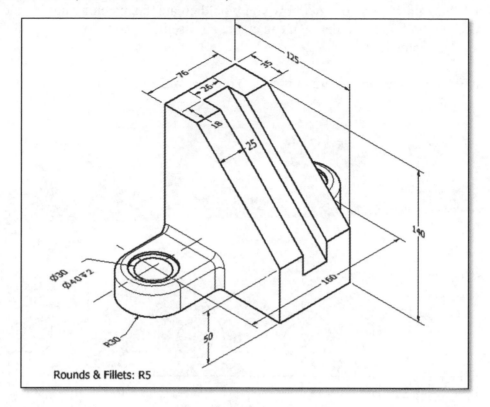

Rounds & Fillets: R5

3. **Piston Cap** (Dimensions are in inches.)

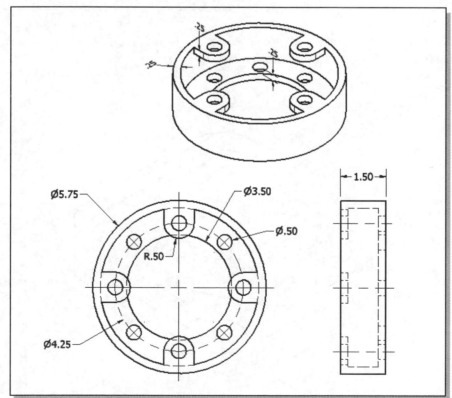

4. Using the same dimensions given in the tutorial, construct the other half of the dryer housing. Plan ahead and consider how you would create the matching half of the design. (Save both parts so that you can create an assembly model once you have completed Chapter 13.)

5. **Intake Flange** (Dimensions are in inches. Thickness: .25 inches.)

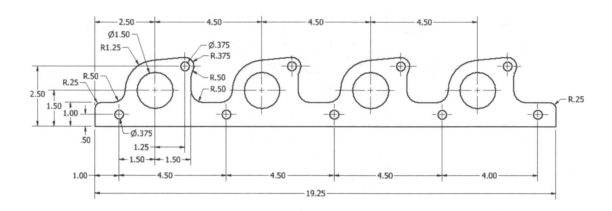

6. **Anchor Base** (Dimensions are in inches.)

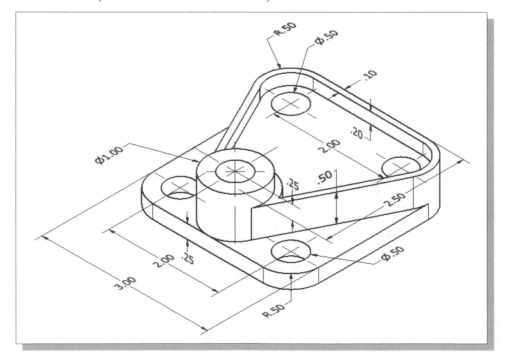

Chapter 13
Assembly Modeling – Joint & Animation

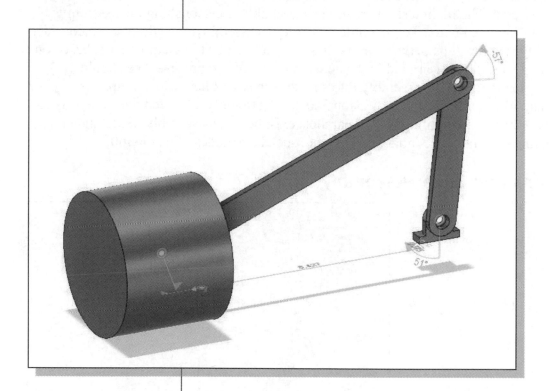

Learning Objectives

- ◆ **Assembly modeling methodology**
- ◆ **Understand the different Assembly Joint type**
- ◆ **Use the Autodesk Fusion 360 Joint Origin**
- ◆ **Use the Animate Model command**

Introduction

In the previous lessons, we have gone over the fundamentals of creating basic parts and drawings. In this lesson, we will examine the assembly modeling functionality of Autodesk Fusion 360. We will start with a demonstration on how to create and modify assembly models. The main task in creating an assembly is establishing the assembly relationships between parts. To assemble parts into an assembly, we will need to consider the assembly relationships between parts. It is a good practice to assemble parts based on the way they would be assembled in the actual manufacturing process. We should also consider breaking down the assembly into smaller subassemblies, which helps the management of parts. In Autodesk Fusion 360, a subassembly is treated the same way as a single part during assembling. Many parallels exist between assembly modeling and part modeling in parametric modeling software such as Autodesk Fusion 360.

The Crank and Slider Assembly

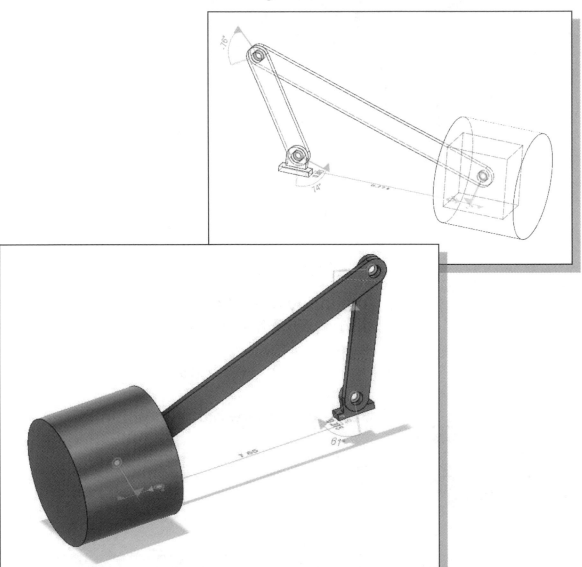

Assembly Modeling Methodology

The Autodesk Fusion 360 assembly modeler provides tools and functions that allow us to create 3D parametric assembly models. An assembly model is a 3D model with any combination of multiple part models. *Parametric assembly constraints* can be used to control relationships between parts in an assembly model.

Autodesk Fusion 360 can work with any of the assembly modeling methodologies:

The Bottom Up Approach
The first step in the *bottom up* assembly modeling approach is to create the individual parts. The parts are then pulled together into an assembly. This approach is typically used for smaller projects with very few team members.

The Top Down Approach
The first step in the *top down* assembly modeling approach is to create the assembly model of the project. Initially, individual parts are represented by names or symbols. The details of the individual parts are added as the project gets further along. This approach is typically used for larger projects or during the conceptual design stage. Members of the project team can then concentrate on the particular section of the project to which he/she is assigned.

The Middle Out Approach
The *middle out* assembly modeling approach is a mixture of the bottom-up and top-down methods. This type of assembly model is usually constructed with most of the parts already created, and additional parts are designed and created using the assembly for construction information. Some requirements are known and some standard components are used, but new designs must also be produced to meet specific objectives. This combined strategy is a very flexible approach for creating assembly models.

The different assembly modeling approaches described above can be used as guidelines to manage design projects. Keep in mind that we can start modeling our assembly using one approach and then switch to a different approach without any problems.

Autodesk Fusion 360's assembly modeling tools allow us to create complex assemblies by using components that are created in separate files or create all parts within one assembly file. In Autodesk Fusion 360, parts are created as bodies and/or components. A component can be a single body or multiple bodies; features and sketches can be modified at any time. The sketches and profiles used to build a 3D Model can be fully or partially constrained. Components can also be created, using other components as references, within the assemblies. The basic concept and procedure of creating assemblies using both Bottom Up and Top Down approaches are discussed in this chapter. The use of the **Joint** command to align and constrain moving components to form an assembly model will be illustrated in the tutorial.

Autodesk Fusion 360 Bodies and Components

In Autodesk Fusion 360, **Bodies** are parametric modeling entities that represent the final shapes of the designs. **Components** are pieces used in an assembly that can be defined to have a certain range of motions. Components are derived from bodies; therefore, a component is a container of bodies. Autodesk Fusion 360 also allows the creation of blank components, as a place holder for future parts to be created later. Both of the bodies and components are identified in the *Browser* area. *Bodies* are essentially collections of three-dimensional features connected to one another, until they form a static representation of a part. Multiple bodies can be contained within one component. In Autodesk Fusion 360, we can be modeling one body relative to the next within the same assembly model; this approach allows users to *borrow* geometry from one body to drive features on the next. It's a very powerful and convenient way to model.

Components can contain one or multiple bodies. Typically, though, as our design takes shape and movements are created within the model, we will only want to see a single body in each component. Components can represent either parts or sub-assemblies, so naturally sub-assembly components will be made up of other components. Note that Fusion 360 allows the creation of components within the assembly model or to insert outside components through the Fusion 360 Data Panel.

The Assembly Joint Command

The **Joint** command is a simple, yet powerful way to position components and define any motions in between components. Joints can only be applied to components; bodies cannot have joints applied. Creating a joint connection allows the user to very quickly define the relations of alignments and movements in between components in an assembly. The *Joint* command can be viewed as creating packaged constraint sets that can be used to allow specific motion (degrees of freedom.) There are seven options available under this command: **Rigid**, **Revolute**, **Slider**, **Cylindrical**, **Pin-Slot**, **Planar** and **Ball** connections.

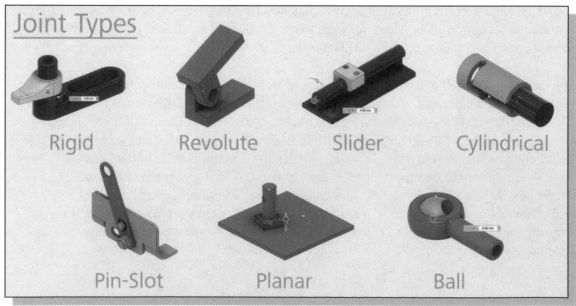

In Autodesk Fusion 360, there are seven different types of *Joints* that can be used with both **As-Built Joint**, and **Joint** commands. Each *Joint Typ*e allows a different number of allowed degrees of freedom (DOF) to define the desired motion. Also, note that with each joint that is created, the associated offset values and parameters can also be altered at any time through the *Timeline Control*. The **Joint** command is used to change the degrees of freedom on components; selecting a *Joint Type* opens the degrees of freedom at the specified translational or rotational directions. Generally speaking, keeping the fewest degrees of freedom open is the fastest way to test and achieve the desired motions in an assembly.

Joint Symbol	Joint Type	Description	Motion Allowed
	Rigid	Locks components together, removing all degrees of freedom.	None
	Revolute	Allows the component to rotate around joint origin.	One rotation
	Slider	Allows the component to translate along a single axis.	One translation
	Cylindrical	Allows the component to rotate and translate along the same axis.	One translation and one rotation
	Pin-slot	The component can rotate about an axis and translate about a different axis.	One translation and one rotation
	Planar	Allows the component to translate along two axes and rotate about a single axis.	Two translation and one rotation
	Ball	Allows the component to rotate about all three axes using a gimbal system (three nested rotations).	Three rotation

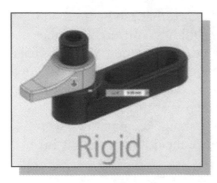

Rigid is probably the most commonly used *Joint Type* at the starting of an assembly. Rigid locks components together, removing all degrees of freedom; for example, when two pieces will be welded together, or when multiple components are bolted together. With the Rigid Joint Type, no additional movement options are available.

The **Revolute** *Joint Type* allows the component to rotate around the **Joint Origin** that has been defined. With the *Revolute Joint*, only one rotation motion is allowed. The *Revolute Joint* is applied when we want to allow rotation about a single point, with no additional translation allowed. Common cases for this are when working with linkages or bolted components rotating about a single point. With the Revolute Joint Type, the *Rotate option* is available to define which axis to rotate about the Joint Origin.

The **Slider** *Joint Type* allows a component to translate along a single axis. With the *Slider Joint*, only one translation motion along an axis is allowed. A common use for this joint is when a component needs to have the ability to move back and forth on a rail without rotation. With the Slider Joint, one Axis selection option is available to define the Slide direction.

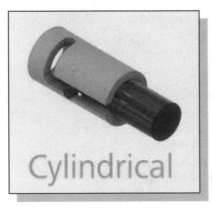

The **Cylindrical** *Joint Type* allows the component to rotate and translate along the same axis. With the Cylindrical Joint, one translational motion and one rotational motion along and about the same Axis are allowed. Cylindrical Joints are great for creating motions on threaded components. With the Cylindrical Joint, the Axis selection option is available.

The **Pin-Slot** *Joint Type* allows the component to rotate about an axis and translate about a different Axis. With the Pin-Slot Joint, one translational motion along an Axis, and another rotational motion around a different Axis are allowed. Note that *Pin-Slot* is very similar in behavior to Cylindrical except the same Axis cannot be used for both motions. With the Pin-Slot Joint, both the Rotate Axis and the Slide Axis selection options are available.

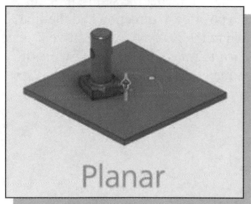

The **Planar** *Joint Type* allows the component to translate along two axes and rotate about a single axis. With the Planar Joint, motions along two translation directions and one rotational direction are allowed. The Planar Joint is great when two components need to be sliding relative to each other on the same plane. With Planar Joints, the Normal direction and the Slide direction selection options are available.

The **Ball** *Joint Type* allows the component to rotate about all three axes using a gimbal system (three nested rotations). The Ball Joint allows rotations about all three axcs, and allows no translational motions. The Ball Joint works great when full rotational freedom is required, but the rotate point should never move. With the Ball Joint, the **Pitch** and **Yaw** options are available.

The *Pitch* option is used to define the **Lateral Axis**, whereas the *Yaw* option is used to define the **Perpendicular Axis**. *Pitch* and *Yaw* are not able to use the same axis; they need to be different. The **Pitch** and **Yaw** options are available to set up the orientation of the expected rotation with the values associated with the direction of the movement. Both Pitch and Yaw can be defined by a specific Axis or by a custom Axis.

The Autodesk Fusion 360 Joint Origin

In Autodesk Fusion 360, a small white disk symbol will be displayed and used on the components when using the **Joint** command. This **Joint Origin** is used to position the components by snapping to pre-generated points. This *Joint Origin* is a visual representation that allows us to understand how the components will be assembled. To apply a Joint relation, we select a *Joint Origin* on each component to assemble the components together. The following figure shows the XYZ directions of the joint origin.

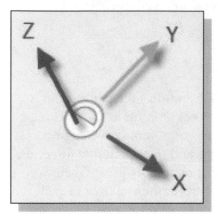

The **Joint Origin** is a coordinate system to show the orientation of alignments of components. The *Joint Origin* plane defines the XY plane; the semi-circle of the *Joint Origin* is the positive Y direction and the line at the center is aligned to the X Axis. The positive Z is perpendicular to the *white side* of the symbol, while the negative side for Z direction is perpendicular to the *light yellow side* of the Joint Origin.

Joint Creation order – "Put This, There."

The **Joint** command is used to set up the relationship in between two components; the selection order will affect the direction of the component's movement. The Autodesk Fusion 360 design team has suggested the users use the phrase "**Put this, there**" as a reminder when creating joint connections in assemblies.

Joint creation

"Put this, there."

- This = selection 1
- There = selection 2

With the **Joint** command we move the first selected component to the location of the second selected component. This allows us to move our components, as well as define the relative motion. Since the first component will move, the first selection will **NOT** allow a selection of a grounded component.

- The first component selected is used to set the movement direction and a blue arrow is displayed in the joint display, relative to the second selection as shown in the figure.

Starting Autodesk Fusion 360

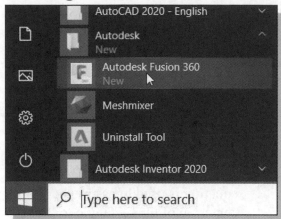

1. Select the **Autodesk Fusion 360** option on the *Start* menu or select the **Autodesk Fusion 360** icon on the desktop to start Autodesk Fusion 360.

2. In the *Sign In* dialog box, log in with your email or username.

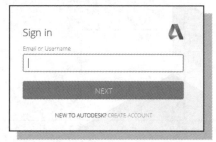

3. Confirm the modeling units set is set to **Inches** in the *Browser* area.

Creating the Base Feature

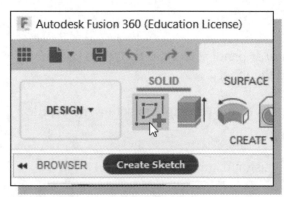

1. Activate the **Create Sketch** icon with a single click of the left-mouse-button.

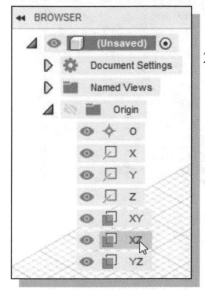

2. Move the cursor on top of the **XZ Plane**, inside the *browser* window as shown, and notice that *Autodesk Fusion 360* will automatically highlight the corresponding plane in the graphics window. Left-click once to select the XZ Plane as the sketching plane.

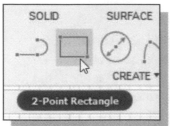

3. Select the **2-point rectangle** command by clicking once with the left-mouse-button on the icon in the *Sketch* toolbar.

4. Create a rectangle of arbitrary size with the center point near the center of the rectangle.

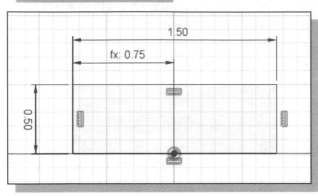

5. On your own, create and modify the dimensions as shown; note the equation is used to assure the centering of the rectangle to the origin.

6. Select **Stop Sketch** in the *Ribbon toolbar* to end the Sketch option.

7. In the *Create Features* panel select the **Extrude** command by left-clicking on the icon.

8. Pick the inside region of the rectangle to set up the profile of the extrusion.

9. Inside the *Extrude* dialog box, set the extrusion distance to **0.25**.

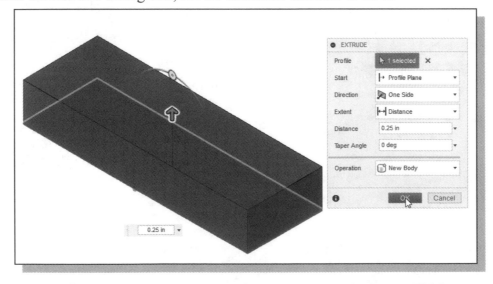

10. Click on the **OK** button to accept the settings and create the base solid feature.

11. Activate the **Create Sketch** icon with a single click of the left-mouse-button.

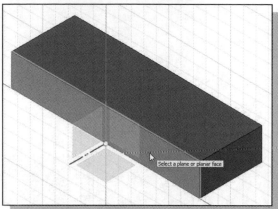

12. Select the front surface of the base feature to align the sketching plane as shown.

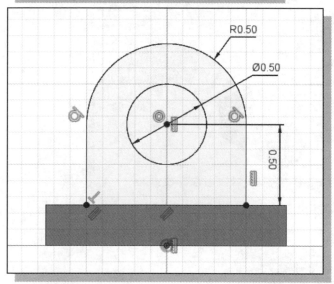

13. On your own, create and modify the sketch as shown; note the center point of the circle is aligned vertically to the origin.

14. Select **Stop Sketch** in the ribbon toolbar area to end the Sketch option.

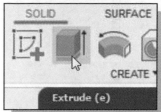

15. In the *Create Features* panel select the **Extrude** command by left-clicking on the icon.

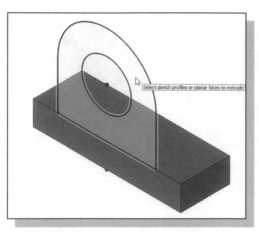

16. Pick the inside region, in between the circle and arc, to set the profile of the extrusion.

17. Drag the arrow toward the backside of the model to set the extrusion direction.

18. Inside the *Extrude* dialog box, set the extrusion distance to **0.25** as shown in the figure.

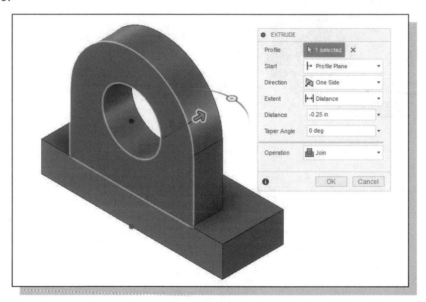

19. Click on the **OK** button to accept the settings and create the first solid model for the *Crank-Slider* assembly.

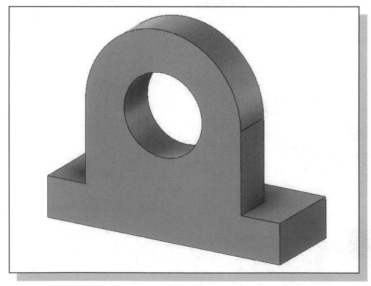

Convert the first Body to the CS-Base Component

In *Autodesk Fusion 360*, components are derived from bodies, therefore a component is a container of bodies. *Bodies* are collections of three-dimensional features connected to one another, until they form a static representation of a part. Multiple bodies can be contained within one component.

1. In the *Browser* area, click on the **Bodies** item to display the item list.

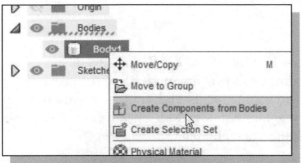

2. Click once with the right-mouse-button on **Body1** to display the option menu.

3. Select **Create Components from Bodies** in the option menu as shown.

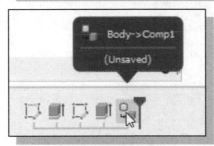

4. Note the last item in the *Timeline Control* indicates the conversion of a **Body** to a **Component**.

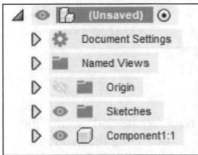

5. Also note that a new item, **component1**, is listed in the browser area. Expand the item list and notice the body is now listed under this new component.

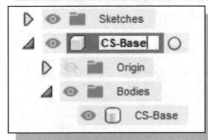

6. On your own, rename both the component and body names to **CS-Base** as shown.

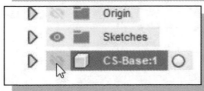

7. Click on the eye icon in front of the CS-Base component to temporarily **hide** the component in the graphics window.

Create the next Component

1. Activate the **Create Sketch** icon with a single click of the left-mouse-button.

2. Move the cursor on top of the **XY Plane**, inside the *browser* window as shown, and notice that Autodesk Fusion 360 will automatically highlight the corresponding plane in the graphics window. Left-click once to select the XY Plane as the sketching plane.

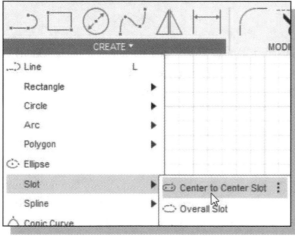

3. Select the **Center to Center Slot** command by clicking once with the left-mouse-button on the icon in the *Sketch* toolbar.

4. On your own, create and modify the sketch as shown; note the center points of the circles are aligned horizontally to the origin.

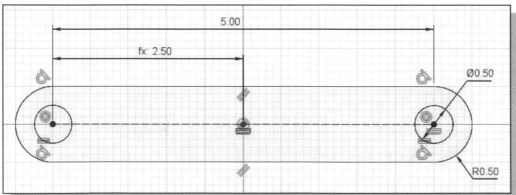

5. Select **Stop Sketch** in the *Ribbon toolbar* to end the Sketch option.

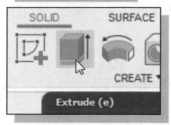

6. In the *Create Features* panel select the **Extrude** command by left clicking on the icon.

7. Pick the inside region, in between the circles and arcs, to set the profile of the extrusion.

8. Inside the *Extrude* dialog box, set the extrusion distance to **0.25** as shown in the figure.

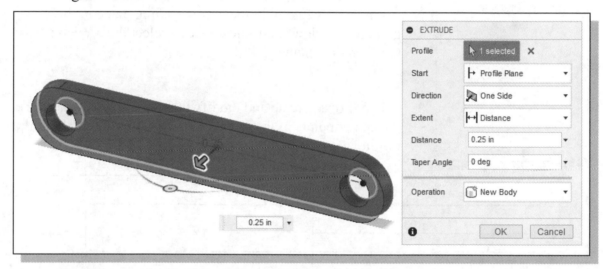

9. Confirm the operation option is set to **New Body** and click **OK** to proceed with creating the new body.

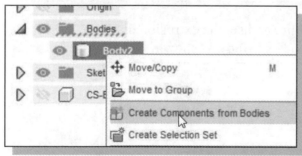

10. On your own, create a new component from the body we just created.

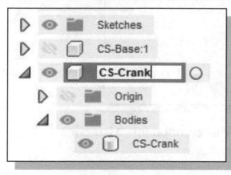

11. Rename the new component to **CS-Crank** as shown.

12. Click on the eye icon and **hide** the component in the graphics window.

Create the CS-Rod Component

1. Activate the **Create Sketch** icon with a single click of the left-mouse-button.

2. Move the cursor on top of the **XY Plane**, inside the *browser* window as shown, and notice that Autodesk Fusion 360 will automatically highlight the corresponding plane in the graphics window. Left-click once to select the XY Plane as the sketching plane.

3. On your own, create and modify the sketch as shown; note the center points of the circles are aligned horizontally to the origin.

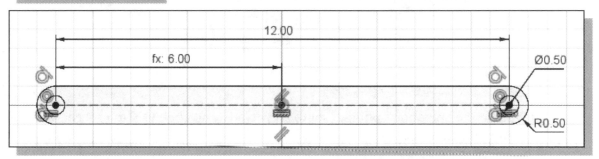

4. On your own, create an *Extrude* feature, and set the extrusion distance to **0.25**. Confirm the operation option is set to **New Body** as shown.

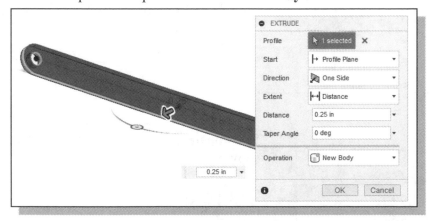

5. On your own, convert the *new body* to the **CS-Rod** component. Also click on the eye icon to hide the component in the graphics window.

Create the CS-Slider Component

1. Activate the **Create Sketch** icon with a single click of the left-mouse-button.

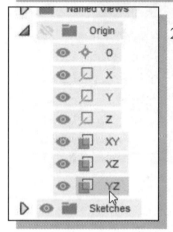

2. Move the cursor on top of the **YZ Plane**, inside the *browser* window as shown, and notice that Autodesk Fusion 360 will automatically highlight the corresponding plane in the graphics window. Left-click once to select the YZ Plane as the sketching plane.

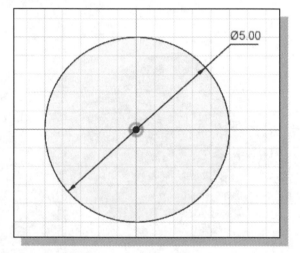

3. On your own, create and modify the sketch as shown; note the center point of the circle is aligned to the origin.

4. Select **Stop Sketch** in the *Ribbon toolbar* to end the Sketch option.

5. On your own, create an *Extrude* feature, set the extrusion distance to **4.0**, and confirm the operation option is set to **New Body** as shown.

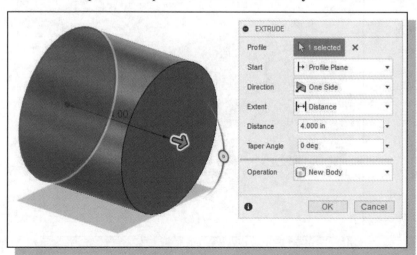

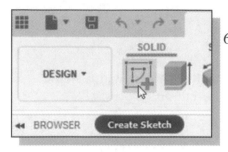

6. Activate the **Create Sketch** icon with a single click of the left-mouse-button.

7. Select the Circular surface as the sketching plane as shown.

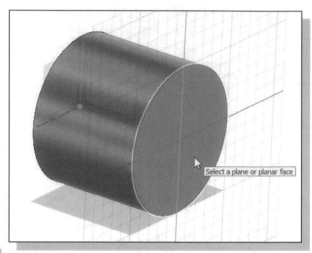

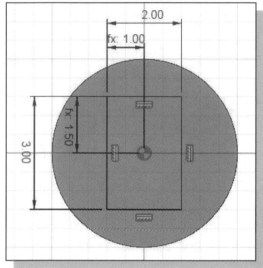

8. On your own, create and modify the sketch as shown; note the center of the rectangle is aligned to the origin.

9. On your own, create an *Extrude Cut* feature, with the extrusion distance to **-3.0** as shown.

10. On your own, convert the *new body* to the **CS-Slider** component.

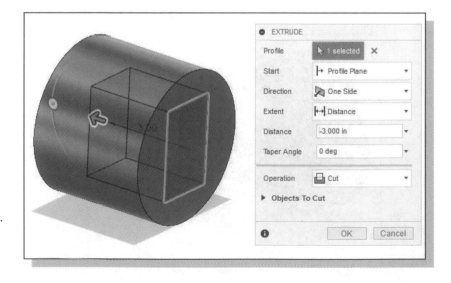

Assemble the First Component

The first component in an assembly file sets the orientation of all subsequent parts and subassemblies and therefore should be one that is **not likely to be removed** and **preferably a non-moving part** in the design. For our project, we will use the *CS-Base* as the base component in the assembly.

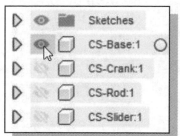

1. In *Browser* area, switch on the **CS-Base Component** command by left-clicking the eye icon.

2. Switch off all of the other components if they are visible in the graphics window.

3. Right-click once to bring up the option menu and select **Ground** as shown. Note that Autodesk Fusion 360 will lock all degrees of freedom for any grounded parts.

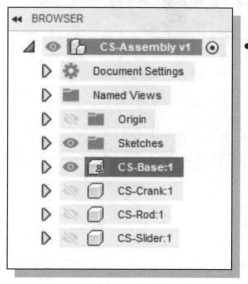

- Inside the *browser* window, the **Pin** icon in front of the CS-Base filename signifies the component is grounded and all *six degrees of freedom* are restricted. The number behind the component name is used to identify the number of copies of the same component in the assembly model.

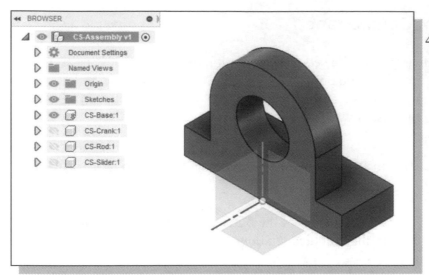

4. In the *Browser* area, switch on the **Assembly Model's Coordinate system** by left-clicking the eye icon.

5. In the *Browser* area, switch on the **CS-Base component's Coordinate system** by left-clicking the eye icon. The CS-Base part has its own coordinate system, which was used when we created the body & component. Both the component coordinates and assembly coordinates are currently aligned at the same location.

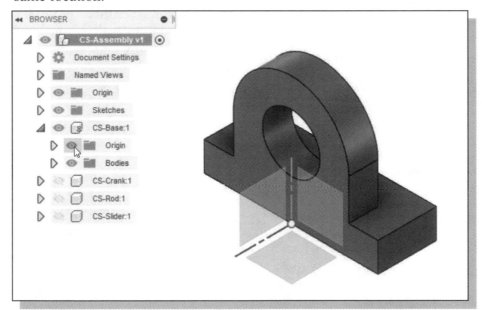

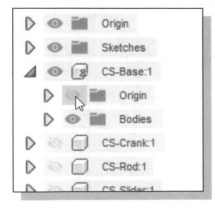

6. On your own, turn off the display of the coordinate systems by clicking on the associated eye icons.

Assemble the Second Component

We will assemble the *CS-Crank* part as the second component to the assembly model.

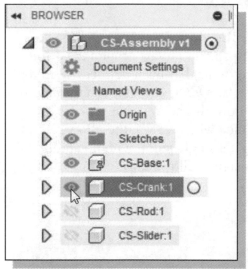

1. In the *Browser* area, switch on the **CS-Crank Component** command by left-clicking the eye icon.

2. In the *Assemble* panel, select the **Joint** command by left-clicking once on the icon.

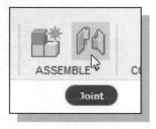

3. Move the cursor near the **hole center** on the left of the CS-Crank component and select the **Joint Origin** aligned to the hole center.

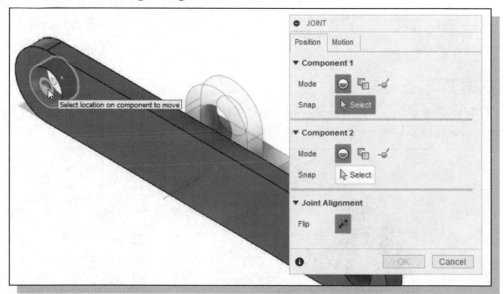

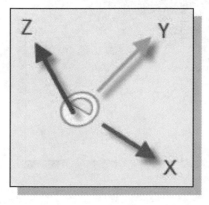

- The **Joint Origin** is a coordinate system designed to show the orientation of alignments of components. The *Joint Origin* plane defines the XY plane, the semi-circle side of the *Joint Origin* is in the positive Y direction and the line at the center is aligned to the X Axis. The positive Z is perpendicular to the *white side* of the symbol, while the negative side for Z direction is perpendicular to the *light yellow side* of the Joint Origin.

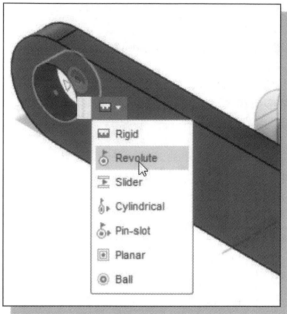

4. In the *Joint* dialog box, set the *Motion Type* to **Revolute** as shown.

- The **Revolute** *Joint Type* allows the component to rotate around the **Joint Origin** that has been defined. With the *Revolute Joint*, only one rotation motion is allowed. The *Revolute Joint* is applied when we want to allow rotation about a single point, with no additional translation allowed.

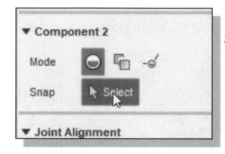

5. In the *Joint dialog box*, activate the **Select** icon in the *Component 2 section* as shown.

6. Click on the front hole center of the CS-Base component to set the *Joint Origin* on the second component and align the two components.

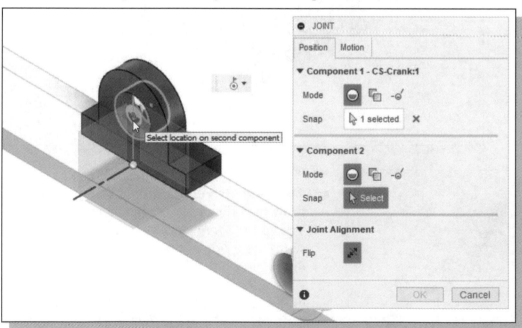

7. In the *Alignment* options, set the angle to **-135 degrees** as shown.

8. Click on the **Flip** icon to set the alignment of the two selected surfaces. Each click on the icon will toggle having the two surfaces facing the same direction or opposite directions.

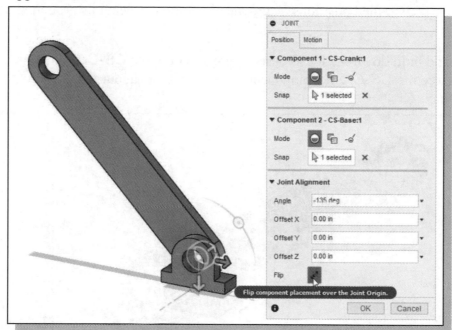

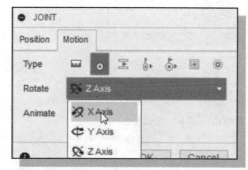

9. In the *Motion* tab, set the *Rotate* option to **X Axis** and observe the animation of the **CS-Crank** component revolving about the CS-Base component.

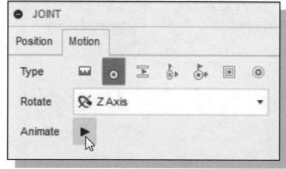

10. In the *Motion* options, reset the *Rotate* option back to **Z Axis.**

11. Click on the **Animate icon** to turn on the animation option.

12. Click on the **Animate icon** again to turn off the animation option.

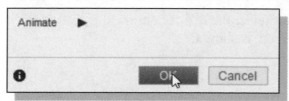

13. In the *Joint* dialog box, click **OK** to accept the joint settings and create the joint connections in between the two components.

Constrained Move

A constrained move is done by dragging the component in the graphics window with the left-mouse-button. A constrained move will honor previously applied assembly constraints. That is, the selected component and parts constrained to the component move together in their constrained positions. A grounded component remains grounded during the move.

1. Press and hold down the left-mouse-button and drag the **CS-Crank** component and notice the *CS-Crank* component can freely rotate about the *CS-Base* component.

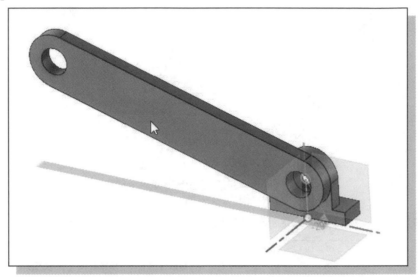

Assemble the CS-Rod Component

We will assemble the *CS-Rod* part as the third component of the assembly model.

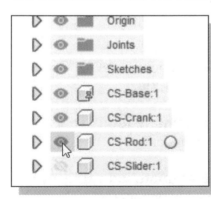

1. In the *Browser* area, switch on the **CS-Rod Component** command by left-clicking the eye icon.

2. In the *Assemble* panel, select the **Joint** command by left-clicking once on the icon as shown.

3. A message dialog box appears on the screen, indicating some components have been moved and two options are available. Note that we did the drag and drop option and moved the CS-Crank component. For our assembly, the new position is not important to our assembly; select **Continue** to reset the position of the CS-Crank component back to where it was prior to the constrained move we did.

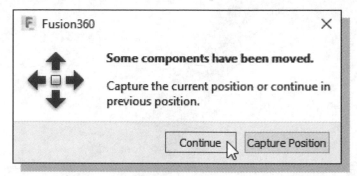

4. Move the cursor near the hole center on the left of the CS-Rod component and select the **Joint Origin** aligned to the hole center as shown.

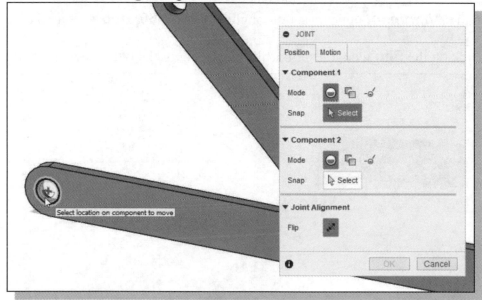

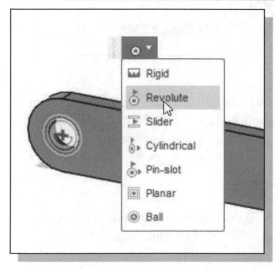

5. On your own, set the *Motion type* to **Revolute** as shown.

6. In the *Joint dialog box*, activate the **Select** icon in the *Component 2 section* as shown.

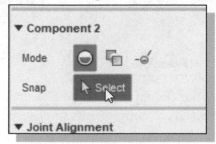

7. Click on the back hole center of the CS-Crank component to set the *Joint Origin* as shown.

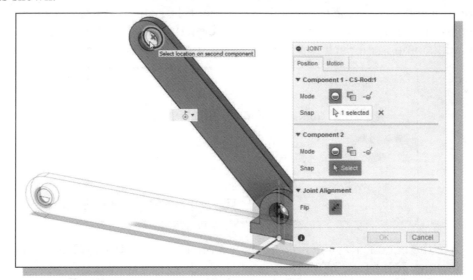

8. In the ***Alignment*** options, set the angle to **60** (or **-60**) **degrees** as shown.

9. Click on the **Flip** icon to examine the alignment of the two selected surfaces.

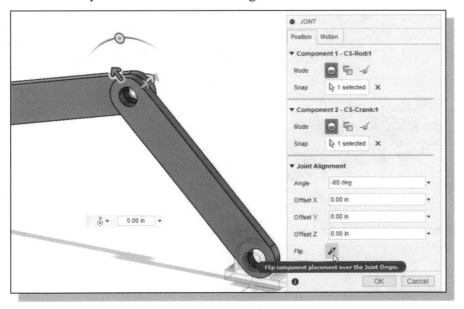

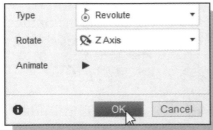

10. In the *Motion tab,* confirm the *Motion type* is set to **Revolute** and *Rotate* about **Z axis**. Click **OK** to accept the joint settings and create the joint connections in between the two selected components.

11. On your own, perform the *Constrained Move* option by dragging and dropping the different components to observe the established assembly constraints.

Assemble the CS-Slider Component

We will assemble the *CS-Slider* component to complete the assembly model.

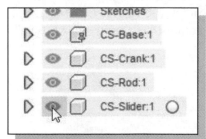

1. In the *Browser* area, switch **on** the **CS-Slider Component** command by left-clicking the eye icon.

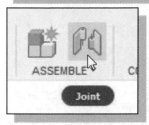

2. In the *Assemble* panel, select the **Joint** command by left-clicking once on the icon as shown.

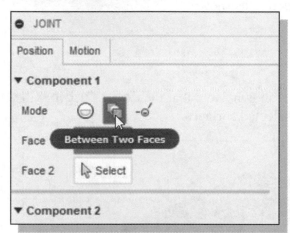

3. In the *Joint dialog box*, select the **between two faces** option in the Component 1 section as shown.

4. Select the two inside vertical surfaces as shown. Note that the **Joint Origin** will be placed on the mid-plane of the two selected surfaces.

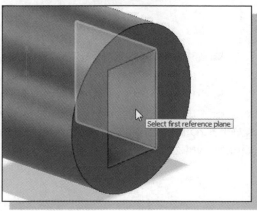

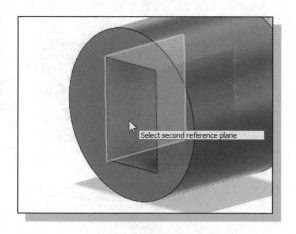

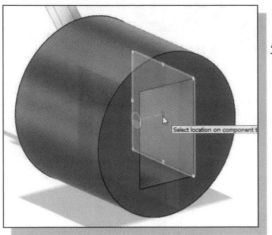

5. Use one of the selected surfaces as reference and place the **Joint Origin** at the center of the rectangular surface as shown.

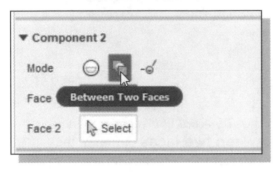

6. In the *Joint dialog box*, select the **between two faces** option in the Component 2 section as shown.

7. Select the two inside vertical surfaces as shown. Note that the **Joint Origin** will be placed on the mid-plane of the two selected surfaces.

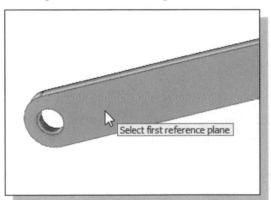
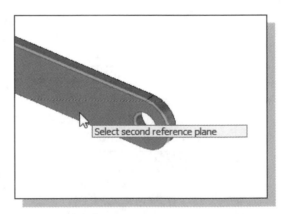

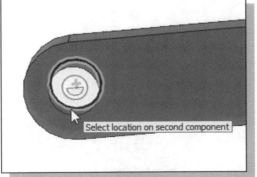

8. Use one of the hole surfaces as reference and place the **Joint Origin** at the center of the circular surface as shown.

9. In the **Alignment** options, set the angle to **-200 degrees** as shown.

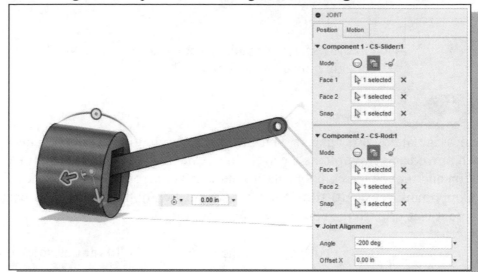

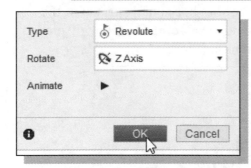

10. In the *Motion* tab, confirm the *Type* is set to **Revolute** and *Rotate* about **Z axis**. Click **OK** to accept the joint settings and create the joint connections in between the two selected components.

11. On your own, perform the **Constrained Move** option by dragging and dropping the different components to observe the established assembly constraints.

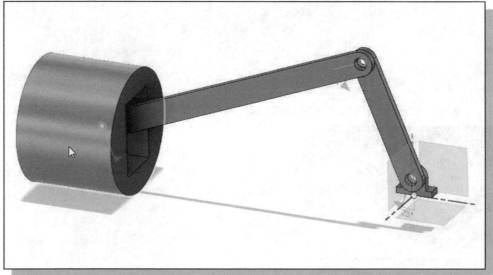

- Note that the motions of the CS-Crank and CS-Rod components are as expected, but the CS-Slider component is still free to spin about the CS-Rod component. We will add another Joint connection so that it can only slide on the horizontal XZ plane.

Apply another Joint Connection

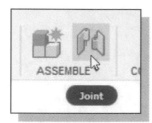

1. In the *Assemble* panel, select the **Joint** command by left-clicking once on the icon as shown.

2. A message dialog box appears on the screen, indicating some components have been moved and two options are available. Note that we did the constrained move option and moved the components around. For our assembly, the new position is not important; select **Continue** to reset the position of the components back to where it was prior to the constrained move we did.

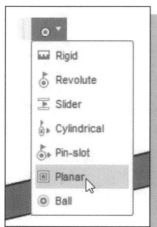

3. On your own, set the motion type to **Planar** as shown.

4. Move the cursor near the bottom horizontal surface of the CS-Slider component and notice a **Joint Origin** is displayed as you move onto a different location.

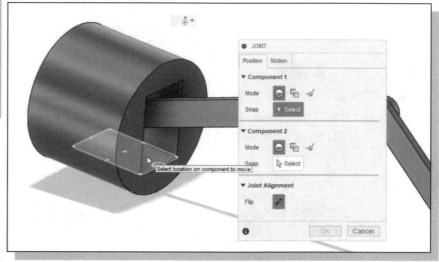

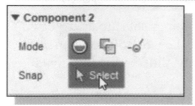

5. Click on the surface to align the **Joint Origin** to the center of the bottom edge as shown.

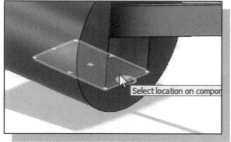

6. In the *Joint dialog box*, activate the **Select** icon in the *Component 2 section* as shown.

7. Move the cursor near the bottom horizontal surface of the CS-Base component and click on the bottom left corner to align the *Joint Origin* as shown.

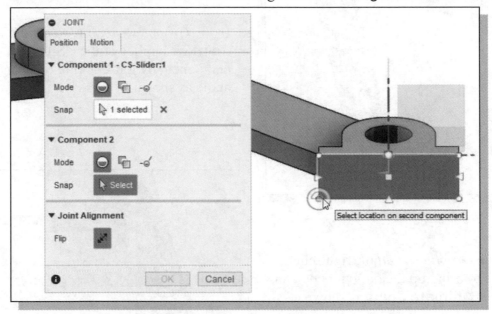

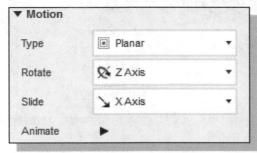

8. In the *Joint* dialog box, set the *Type* to **Planar** and *Normal* to **Z axis** and slide along **X axis**. Click **OK** to accept the joint settings and create the joint connections in between the two selected components.

9. On your own, perform the **Constrained Move** option by dragging and dropping the different components to observe the motion of the assembly as expected.

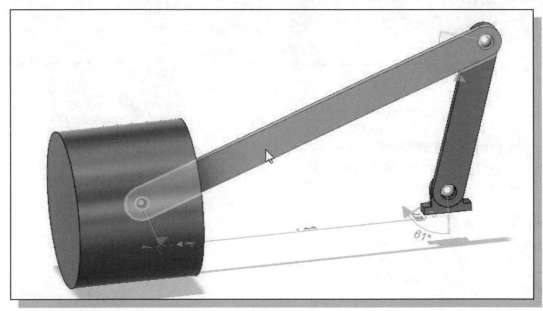

Animate the Assembly Model

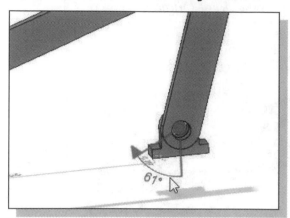

1. In the *Graphics window*, select the first **Joint** we applied at the CS-Crank component by left-clicking once on the angle as shown.

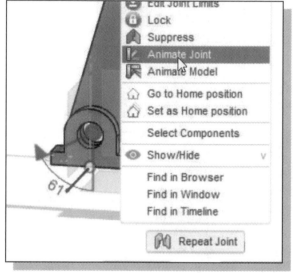

2. In the *Graphics window*, right-click once to bring up the option menu and select **Animate Joint** as shown.

•   Note only the CS-Crank is rotating about the CS-Base component at the aligned location.

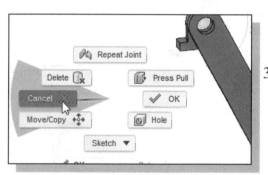

3. In the *Graphics window*, right-click once to bring up the option menu and select **Cancel** to end the animation.

4. In the *Graphics window*, right-click once to bring up the option menu and **Animate Model** in the option menu to start the animation of the entire assembly model.

•   The animation of the assembly model indicates the joint connections are applied correctly showing the proper motion of the assembly.

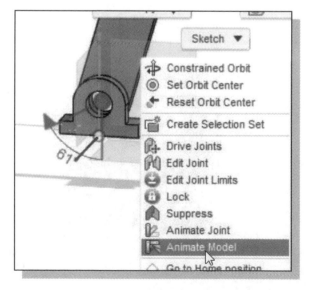

Review Questions:

1. What is the purpose of using *assembly models*?

2. What are the differences between bodies and components in Autodesk Fusion 360?

3. List and describe three types of the Autodesk fusion 360 *assembly Joints*.

4. How should we determine the assembly order of different parts in an assembly model?

5. Describe the procedure to animate the assembly in Autodesk Fusion 360.

6. Describe the procedure to use the **Joint command** in Autodesk Fusion 360.

7. Create sketches showing the steps you plan to use to create the four parts required for the assembly shown on the next page:

Ex.1)

Ex.2)

Ex.3)

Ex.4)

Exercises:

1. **Leveling Assembly** (Create a sct of detail and assembly drawings. All dimensions are in mm.)

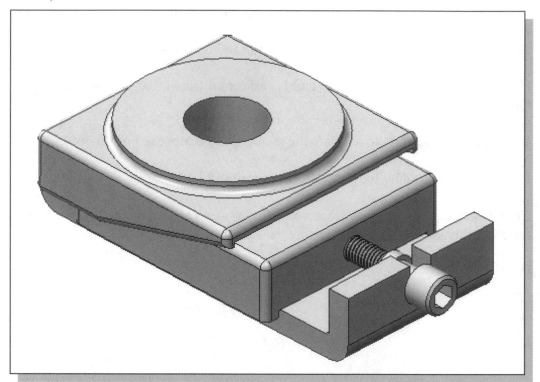

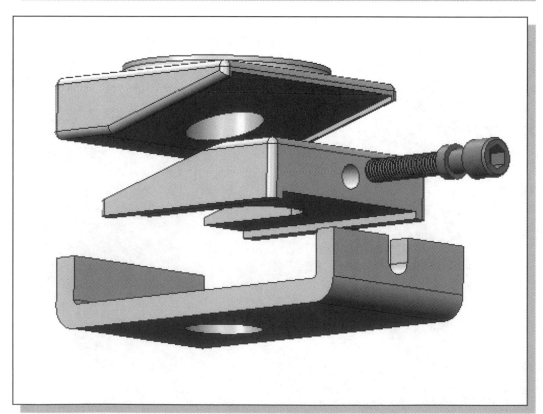

(a) **Base Plate**

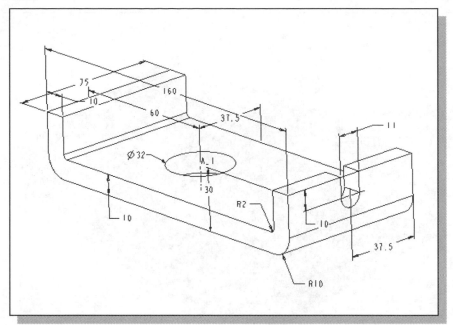

(b) **Sliding Block** (Rounds & Fillets: R3)

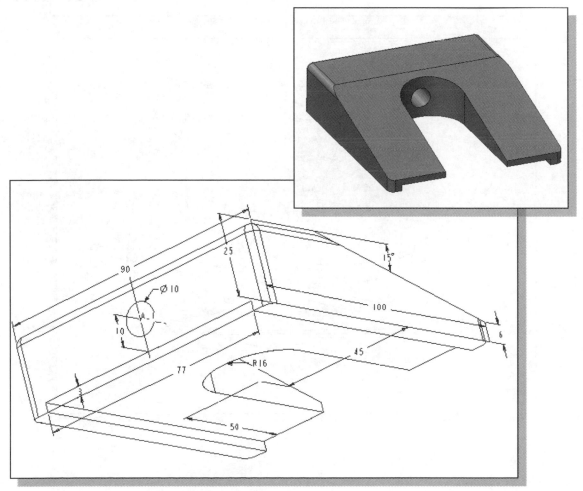

(c) **Lifting Block** (Rounds & Fillets: R3)

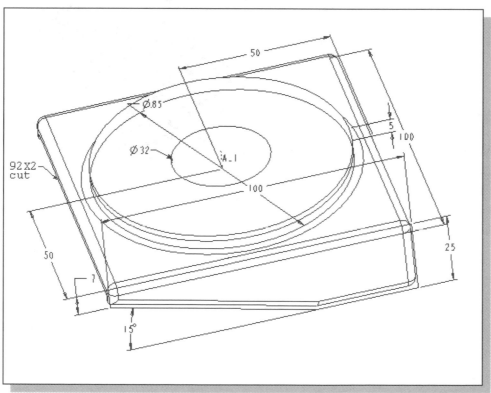

(d) **Adjusting Screw** (M10 × 1.5)

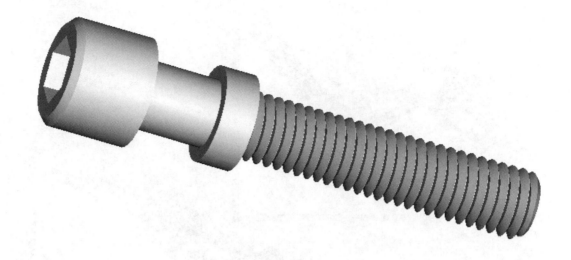

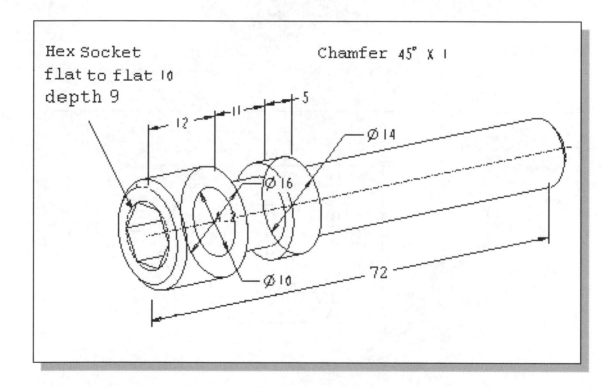

Basic Motion Analysis: Create a linear dimension between the **Sliding Block** and the **Base Plate** parts. Use the *Drive constraint* command to start the animation and perform the interference analysis.

2. Toggle-Clamp Assembly

(Create a set of detail and assembly drawings. All dimensions are in inches.)

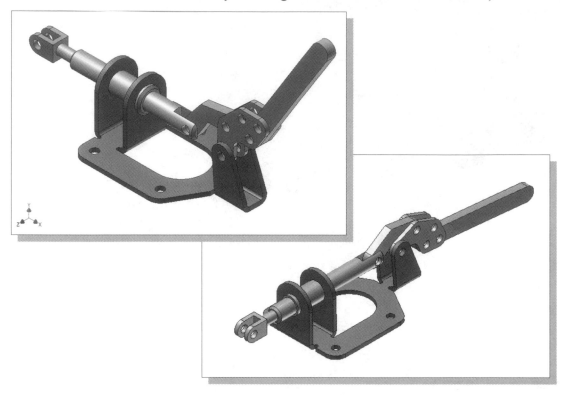

a) Sheet Metal Base

1. No. 11 Gauge (0.125) Mild Steel
2. All Bend Angles are 90 degrees
3. Bend Radius: .5 Thickness
4. Flat Layout K-Factor: 0.40
5. Standard Obround Relief

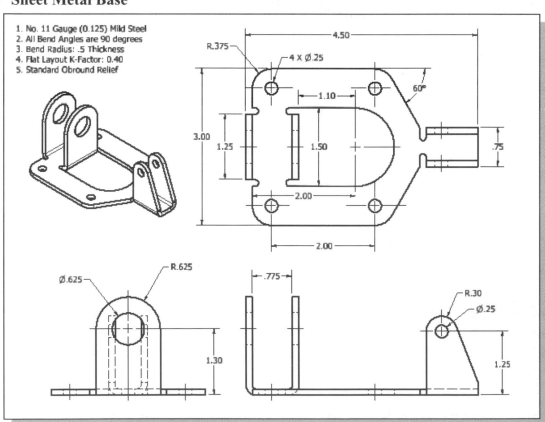

b) Connector

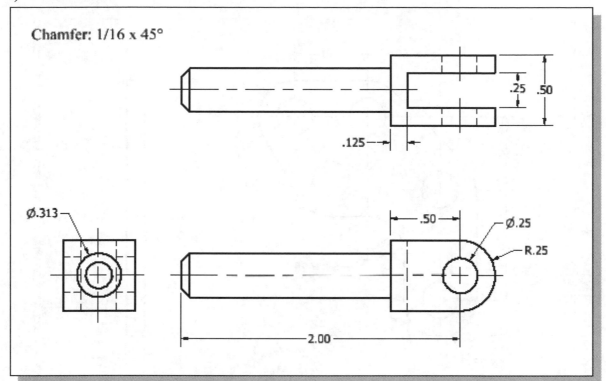

Chamfer: 1/16 x 45°

c) Handle

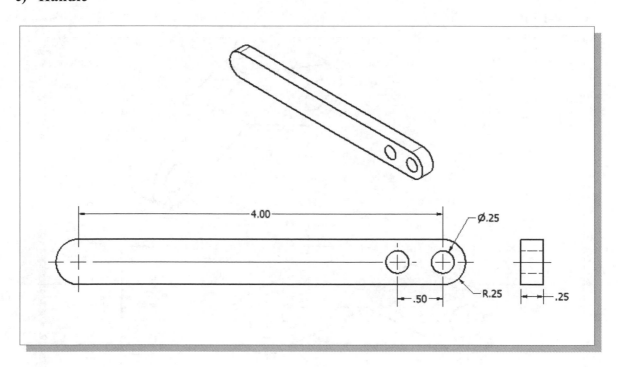

d) Joint Plate

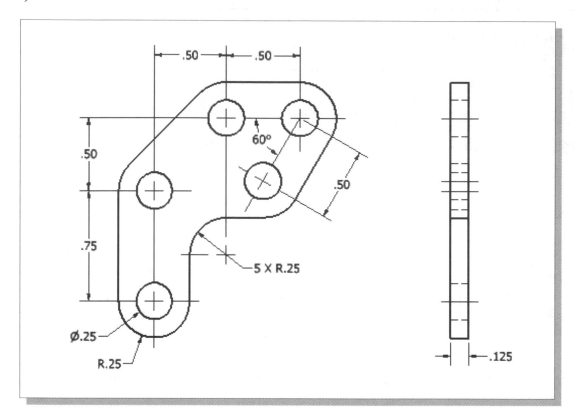

e) V-Link

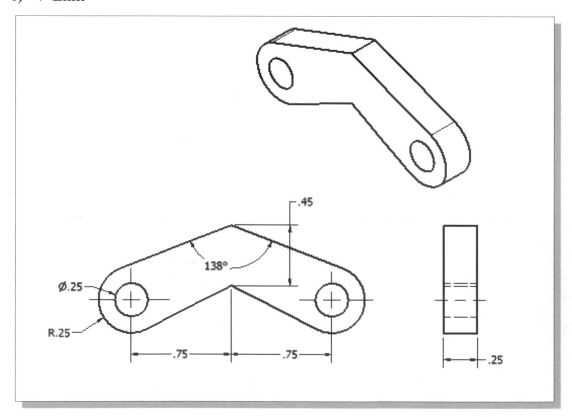

f) Rod

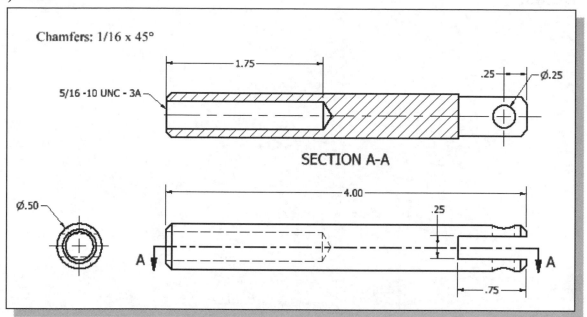

Chamfers: 1/16 x 45°

5/16 -10 UNC - 3A

1.75

.25

Ø.25

SECTION A-A

Ø.50

4.00

.25

.75

A ← A

g) Bushing

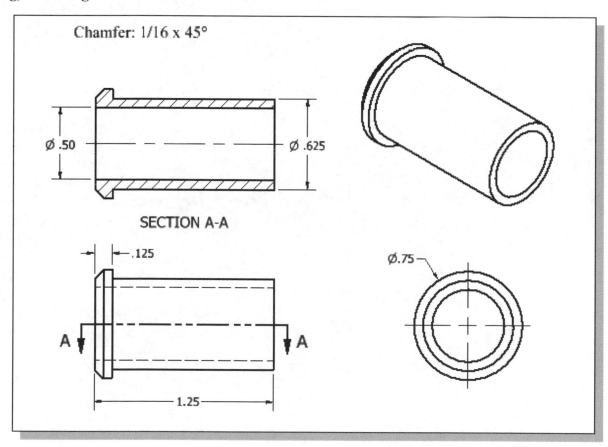

Chamfer: 1/16 x 45°

Ø .50

Ø .625

SECTION A-A

.125

A ← A

Ø.75

1.25

Notes:

INDEX